Science Success for Students with Disabilities

Robert A. Weisgerber, Ed.D.

Senior Research Fellow

American Institutes for Research

Innovative Learning Publications

Addison-Wesley Publishing Company
Menlo Park, California • Reading, Massachusetts • New York
Don Mills, Ontario • Wokingham, England • Amsterdam • Bonn
Paris • Milan • Madrid • Sydney • Singapore • Tokyo
Seoul • Taipei • Mexico City • San Juan

This material is based on work supported by the National Science Foundation under Grant No. TPE-915403. The Government has certain rights in this material. Any opinions, findings, conclusions, or recommendations expressed in this material are those of the author and do not necessarily reflect the views of the National Science Foundation.

Many of the designations used by manufacturers and sellers to distinguish their products are claimed as trademarks. Where those designations appear in this book, and Addison-Wesley was aware of a trademark claim, the designations have been printed in initial caps.

This book is published by Innovative Learning Publications™, an imprint of the Alternative Publishing Group of Addison-Wesley Publishing Company.

Project Editor: Mali Apple
Design Manager: John F. Kelly
Production: Claire Flaherty
Text and Cover Design: Side by Side Studios
Illustrations: Leslie Dunlap
Cover Illustration: D. J. Simison

This Book Is Printed
on Recycled Paper

ISBN 0-201-49089-7
1 2 3 4 5 6 7 8 9 10-ML-98 97 96 95 94

Contents

] *The Key to Alternative Strategies, on page iv, lists the topics included under each of these three sections.*

KEY TO THE ALTERNATIVE STRATEGIES

HI = Hearing Impairment PI = Physical Impairment VI = Vision Impairment
LD = Learning Disability SI = Speech Impairment ED = Emotional Disorder

Symbol Key

K–12   Describes the applicable span of grades for a strategy.

Indicates an activity that takes place in the classroom.

Indicates an activity that takes place in the science center or lab.

Indicates an activity that takes place outside of school.

Acknowledgments

Science Success for Students with Disabilities is the outcome of two research and development studies. The first was a research project funded by the National Science Foundation entitled "Research to Identify Critical Factors Contributing to Entry and Advancement in Science, Mathematics, and Engineering Fields by Disabled Persons" (Grant MDR-8751195). That study clearly revealed the crucial role science teachers play in encouraging young persons with disabilities to pursue higher education and careers, and ultimately reach their potential, in the sciences. The study findings were reported in *The Challenged Scientists: Disabilities and the Triumph of Excellence.* (Weisgerber and Praeger 1991) The second study, also supported by NSF (Grant TPE-915403), focused on developing instructional materials that would be useful to science teachers as they encounter intellectually able students with disabilities in their regular classes.

We consider ourselves fortunate to have had counsel from a uniquely qualified panel of consultants knowledgeable about science, teaching, and learning. Our consultants spoke to the issues and challenges faced by persons with disabilities as they encounter numerous barriers and work out coping strategies. We are indebted to the following Advisors: Mr. Peter Axelson of Beneficial Designs, Inc.; Dr. Ilan Chabay of the New Curiosity Shop, Inc.; Ms. Linda De Lucchi of the Lawrence Hall of Science; Dr. Anne Barrett Swanson of the College of St. Catherine; and Dr. Lynda West of George Washington University.

We also received important help in the design stage from six teachers, representing four school districts. They were Susan Hamada, Verna McDonald, Marla Deutsch, Jean Gallagher, Terry Augustine, and Geri Horsma. Geri, a science teacher in Palo Alto, California, not only helped in the design stage but continued throughout the project as a technical advisor. She proved to be a master teacher as she helped conduct each of the inservice workshops and preservice classes that occurred in the second year of the study. She is a joy to work with.

Phyllis DuBois, a valued colleague at AIR, kept me on track throughout the project. In many respects she deserves the credit for the background work that led to Chapter 5. Her editorial skills proved to be invaluable as the book emerged. My thanks for her contributions.

Finally, I want to express my appreciation to the staff at Addison-Wesley Publishing Company: Michael Kane, Managing Editor, who first recognized the value of a book on this topic; Pat Brill, Editorial Director, who provided encouragement for the second edition; and Mali Apple, Senior Editor, who saw this revised edition through to completion.

R.A.W.

Preface

In developing these materials we have drawn heavily upon the expertise of teachers. The teacher-advisors consisted of two elementary teachers, two middle school teachers, and two secondary teachers. Representing four school districts overall, each teacher was selected for dual experience in teaching science and teaching students with disabilities. The teacher-advisors developed preliminary versions of strategy sheets and later reviewed and amplified draft materials prior to full field testing and publication. The panel helped design a structure and format that would meet several criteria:

- It should be easy to use so teachers can readily access needed information.

- It should reflect good practice in teaching children with specific disabilities.

- It should encourage teacher creativity, flexibility, and expanded thinking.

- It should reflect what is known about effective science teaching and learning.

- Strategies and approaches should be varied to reflect students' disabilities.

Field testing of *Science Success for Students with Disabilities* took place in California, the District of Columbia, Georgia, Minnesota, North Dakota, and Virginia, and included five preservice teacher groups and four inservice teacher groups. All told, 124 teachers and 107 teacher trainees reviewed and provided comments on *Science Success*. The book was uniformly well received, and the second edition reflects refinements and additions suggested by teachers and teacher trainees.

Teachers who use *Science Success for Students with Disabilities* should be aware of these strongly held beliefs of the author:

- Students with disabilities *can* benefit from participation in science classes.

- Serving their needs is often much simpler than one might expect.

- They can be an asset to the class and make cooperative learning meaningful.

- There is no substitute for a positive attitude and a helpful, inspiring, mentoring role.

- Persons with disabilities can enter and succeed in many different science careers.

Hopefully, *Science Success for Students with Disabilities* will help you as you help these students meet the challenge ahead.

Overview

Educators are challenged to help each and every student develop to his or her full potential. This is true academically and socially, and it is true with respect to the fulfillment of personal goals and interests. It is a part of your job, an important part, to learn how you can best support students with disabilities as they develop skills of independence and self-reliance. It can be enormously satisfying when a student with a disability responds in your class, advances toward goals, and clearly benefits from the coursework and the opportunity to learn in a supportive environment that you have fostered.

This guide will help you to become a more effective teacher as you interact with students with disabilities in your science class.

ON HAVING A DISABILITY

It's impossible to put yourself in another person's place, yet you can try to imagine some of the difficulties you would encounter if you went through one school day with a disability. Mental role playing is just a "taste" (but not a diet) of living with a disability–it is merely a starting point for becoming more aware of the needs of persons with disabilities as they go about their daily living. It is also a way to become sensitive to the different barriers and problems that often arise in instructional situations.

The first step to dealing with a problem is becoming aware of it. Take a few moments to try the following Exercises. Consider the questions from the perspective of four different students:

- A student with a hearing disability

- A student with a physical disability, using a wheelchair

- A student who is functionally blind

- A student with a learning disability

Exercise 1: How did your day begin?

First, choose one disability and assume that role in your mind. List the special problems that arise as you go through your morning routine. Summarize the *extra* daily tasks you, the student, must do to stay abreast. Repeat the exercise for each of the other disabilities or reflect on what might happen differently.

Exercise 2: What happened to you in science class today?

Below are four situations involving a hands-on activity that might occur in a science class. Imagine that the teacher has made no special adjustments for you, the student with a disability. Reflect for a few moments on what you would learn in each lesson, then answer the questions that follow.

- *You are a student with a hearing disability.* The science lesson is about sound. The current lesson plan calls for using glasses partially filled with water and a metal spoon to tap the glasses in an activity in which teams of students are to develop an array of tones for a recognizable version of "Jingle Bells" and explain the key principles involved.

- *You are a student with a physical disability.* The science lesson is about water quality. The current lesson plan calls for students to collect water samples in the neighborhood (other than from faucets), "grow" organisms in the samples, examine the specimens with a microscope, and describe their observations using sketches and counts.

- *You are a student who is functionally blind.* The science lesson is about chemical interaction. The current lesson plan calls for using baking soda and vinegar in specific proportions and charting the reactions with different amounts of each chemical.

- *You are a student with a learning disability.* The science lesson is about telling time. The current lesson plan calls for constructing a simple sundial, placing it outside at noon, placing marks on the sundial at hour intervals, and reporting differences in shadow length over a week's time.

What barriers would you, as the student with a disability, be likely to encounter in each situation? In framing your answer, consider the following:

What instructions did the teacher give? Were you able to understand them?

What tasks were the other students given? What tasks were you given?

What happened when you tried to participate?

What barriers to learning could have been reduced or eliminated?

What might your teacher have done differently?

What might your teacher have done to facilitate the lesson and better support you?

How would the teacher know that you were mastering the topic?

As a result of this mental exercise, it should be apparent to you that (a) students with disabilities encounter many barriers throughout the day and in the classroom, (b) different students have different kinds of problems, and (c) the teacher plays a key role and what he or she does can make a big difference in whether the student with a disability participates fully and functionally in science class.

This guide will provide you with ideas you can use in planning your instructional program to benefit these students.

HOW TO USE THIS GUIDE

Science Success is neither a quick cure nor a panacea. It will not make problems disappear, but it will help you to address problems (we prefer to think of them as challenges) in a way that minimizes their impact. It will also give you

- an appreciation of your role when intellectually able students with disabilities are placed in regular science classes

- ideas to help enable you to be a teacher, friend, and mentor to a student with a disability

- suggestions for instructional strategies that have proven successful for others

- resources you can turn to for help and counsel

Chapter 1 provides a background of understanding about current trends in science teaching and an appreciation for the potential of students with disabilities.

Chapter 2 provides a core of knowledge about different disabilities that will be helpful as you plan instructional experiences.

Chapter 3 will alert you to the three types of barriers that students with disabilities often encounter. It suggests a number of constructive steps you can take to help students with disabilities improve their performance.

Chapter 4 is a comprehensive survey of barriers that occur in (a) teacher-centered instruction, (b) student-centered activity, and (c) partner- and team-based learning. It also provides numerous suggestions for alternative instructional strategies that can benefit students with different disabilities.

In Chapter 4, headings are used as a way of quickly orienting you to page contents. The heading shown below can be interpreted as follows:

Section A	**hearing impairment**
Teacher-Centered Instruction	**physical impairment**
	vision impairment
USING MULTIPLE MODALITIES	**learning disability**
	speech impairment
	emotional disorder

The *section* is Teacher-Centered Instruction.

The *topic* is Using Multiple Modalities.

Only four of the six disabilities (highlighted with bars) have a strategy page, and these pages follow in the order of the bars. Strategy pages are not provided where it is felt that the disability should not prevent the student from functioning within the range of performance expected of nondisabled students.

Chapter 5 points you toward additional resources from around the country. These resources can be accessed to better support students with disabilities in the science program at your school.

Appendix A is designed to help school districts organize inservice workshops so that science teachers, special education teachers, and other resource specialists in the schools can develop a *coordinated* approach to the instruction of students with disabilities in the regular science program.

Appendix B gives specific suggestions for introducing the topic of disability in preservice classes in science education at the college and university level. Students in education methods classes and those engaged in student teaching can benefit from knowing how they can interact effectively with students with disabilities in their classes.

Appendix C provides vignettes to foster individual or group consideration of specific cases. They include questions to focus discussion and to motivate referral to the various chapters of *Science Success*.

Recognizing Potential in Students with Disabilities

There are special things about each one of us. We all have the potential to be something more than we are today—we can grow and thrive as we reach to fulfill our potential. We are all limited in some ways. Frequently, we become frustrated when our skills are limited and we are not given the opportunity to develop those skills and fulfill our potential. This book is about helping elementary, middle school, and secondary students realize their potential in science. More specifically, it is about deciding whether you, as a teacher, are willing to provide opportunities and reduce barriers so that *intellectually able students with disabilities* can benefit from being a part of any science-oriented class you teach.

This guide is about helping students reach their potential.

THE VIEWPOINT IN THIS GUIDE

Science Success is a guide for regular classroom teachers who are teaching science to nondisabled children and who have one or more students with disabilities in their class. Because we are primarily concerned with students who can benefit from being mainstreamed, we will focus on those disabilities in which intellectual capacity is not (ordinarily) impaired. We will discuss instructional approaches for students with physical, visual, and hearing impairments, as well as emotional disorders, speech impairments, and learning disabilities.

The contributors for *Science Success* are able to say, from personal experience, that students with disabilities can benefit from science classes, develop a flair for and interest in science, and become successful in science careers. We know many individuals who have realized their potential in science even though they are disabled in some way.[1,*]

Many of these students were able to accomplish this because someone took an interest in them and provided the constructive support they needed. At the same time, we are keenly aware that some of them accomplished their personal goals even though they were actively discouraged from pursuing science careers by teachers who simply did not know how to give them the support they needed and didn't believe they could function in science at advanced levels.

> In the past, little attention was given to involving students with disabilities in regular science classes. We can produce more opportunities and a more productive future in the sciences for students with disabilities.

CREATING A SCIENCE-LITERATE CITIZENRY

The future of science in the United States depends on our finding and nurturing science interest and talent, and on our ability to create a science-literate citizenry. That can only occur if skilled science teachers make it happen.

A strong case can be made that effective science teaching is a skill that arises from a love of the subject matter, a respect for the students, and a willingness to experiment with various instructional strategies. Science teachers are frequently resourceful and inventive in conveying their subject matter to their students. Teachers are finding it especially rewarding to involve students actively in

> Science teaching involves being willing to experiment with various instructional strategies to meet the needs of different students.

- hands-on learning

- cooperative learning in student teams

- multisensory classroom, laboratory, and outdoor experiences

INCLUDING STUDENTS WITH DISABILITIES

Quite simply, you should keep in mind that people with disabilities also have *abilities*. More and more of these individuals are seeing their disability as an inconvenience, not as something that prevents them from doing things.[2] You can deal best with the student's disability when you focus on

> Focus on the abilities—not the disabilities—of the students.

- his or her abilities

- reducing attitudinal and physical barriers

- a variety of instructional techniques that provide lots of positive support and encouragement

---

*Notes appear at the end of the chapter.

Goals for teaching students with disabilities

To reach your potential as a teacher of students with disabilities, be prepared to make an investment. Here's what you need to do.

Make it a personal goal to

- accept students with disabilities into the classroom with a positive attitude
- set reasonable goals together
- involve students with disabilities in classroom activities
- engage nondisabled students in productive interaction with disabled students

Engage all the students in productive interaction directed toward attainable goals.

Goals for teaching science classes

Approach science teaching as you would have liked it to have been taught to you.

- Make it fun: let the students share your interest and build on their experiences.
- Present science as one way to gather evidence and draw good conclusions.
- Present science as a way to seek answers to the questions: Will? How? Why?

Make science interesting and show that it can be a good way to answer questions that arise in life.

APPROACHES TO SCIENCE EDUCATION

Science Success is not intended to teach you how to teach science. But we do suggest that instructional approaches that are effective for all students involve

- hands-on, multisensory learning
- cooperative learning
- interdisciplinary learning

More will be said about these approaches later, so they will only be summarized here.

Hands-on learning

Elementary and secondary teachers increasingly recognize the benefits of building opportunities for multisensory and hands-on experiences into the curriculum to meet the diverse learning needs and styles of students and to enrich their educational experience. What does *hands-on* really mean? Scientists and engineers have pointed out that it does not simply mean "handle the goods"; it means "control the operation."

In other words, it is less important that the student personally manipulate objects than it is for the student to effectively direct the handling of the objects by another person. Understanding the concepts that underlie the procedures is more important than simply executing the mechanics of the procedures. Accordingly, students with disabilities should participate in such hands-on activities as much as possible. In no case should they be excluded simply because they have difficulty handling the materials.

Students with disabilities may require an adaptive approach to learning to address their specific challenges. Whether a student's disability is hearing, vision, physical, emotional, speech centered, or related to the learning process itself, teachers must design and use strategies that enhance the student's ability to experience, comprehend, and achieve in the school environment.

Hands-on learning means much more than "handle the goods"—it means "control the operation."

Hands-on learning often provides multisensory experiences.

A *multisensory* approach to learning is particularly well suited to science education. It provides rich opportunities for teaching students with disabilities in diverse learning situations, including mainstreamed classrooms. The Multisensory Learning Center at the University of California-Berkeley's Lawrence Hall of Science has been successful with this approach—first in developing materials to teach science to blind and visually impaired students and later in adapting materials for use with students with diverse disabilities and students in regular education classes.

Cooperative learning

In a regular science classroom, disparities in the past experiences and present capabilities of students may be great. Generally, students with disabilities have had less prior experience and fewer opportunities to explore on their own. Instead of treating diversity as a barrier, teachers should recognize that these differences can actually contribute to classroom experiences shared by all students. Students with disabilities can contribute.

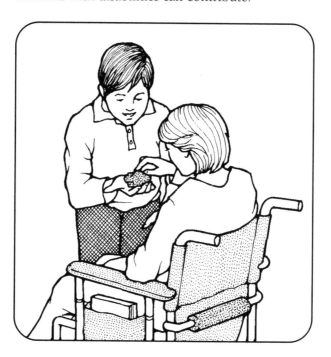

In cooperative learning, students in groups or pairs can share their respective skills.

In cooperative learning, students with disabilities are teamed with other students who can provide the hands, feet, eyes, and ears for them. Such teaming can be critically important to a student with a disability, as long as the peers value the abilities and are sensitive to the limitations of the student they are assisting and do not overshadow or patronize.

Interdisciplinary learning

Various national planning groups in science education have made a convincing case for an interdisciplinary approach to the teaching of science. Science is not isolated from reading, writing, and mathematics. Science projects and assignments typically involve elements of all these disciplines.

Science learning involves other school subjects.

Students with disabilities sometimes find it helpful to approach a science topic from the different perspectives of reading, writing, and mathematics in addition to the hands-on approach that typifies much of science education. The interdisciplinary approach provides reinforcement, and it offers students with disabilities more opportunities to make real contributions in the class.

SPECIAL EDUCATION COOPERATION

Regular teachers and special education teachers can work together in meeting the needs of the student with disabilities. The special education teacher has had experience in classroom communication that minimizes barriers to learning for students with disabilities. Special education staff generally welcome the opportunity to share their knowledge of techniques with regular teachers. They can also share information about the disability and its effect on each student.

The special education teacher and the regular (science) teacher can work together to share their knowledge of techniques and strategies.

On the other hand, special education teachers typically know little about the content of science teaching and need to be familiarized with the instructional goals the regular (science) teacher has set. Early cooperation and planning between regular and special teachers can be an enormous help in anticipating and smoothing out problems before they arise.

Early and frequent collaboration between regular and special educators is important.

LEARNING ABOUT DISABILITIES

By law, schools are required to document carefully the present performance, strengths, and weaknesses of the students with disabilities they serve and to provide appropriate goals for these students so they can participate in and benefit from education to the maximum extent.

For various state and federal reporting purposes, and as a product of the diagnostic process in which their special needs are identified, students' disabilities are classified into categories. The categories used by the Department of Education in its report to Congress, consistent with the Individuals with Disabilities Education Act (IDEA, P.L. 101-476), are listed below. However, states can and do differ in terms of definitions and classification criteria applied.

Disability "labels" arise from diagnosis and a need for reporting.

About 4,260,000 students with disabilities are served in the schools.

Categories of disabilities used by the Office of Special Education Programs, Department of Education, together with student counts, as of May 1991:

Specific learning disabilities	2,064,892
Speech/language impairments	976,186
Mental retardation	566,150
Serious emotional disturbance	382,570
Multiple disabilities	87,956
Hearing impairments	58,164
Other health impairments	53,165
Orthopedic impairments	47,999
Visual impairments	22,960
Deaf-blindness	1,634

The definition of disability in the Americans with Disabilities Act (ADA, P.L. 101-336) is broader than the one used by the Department of Education. For one thing, it considers disability in terms of functionality.

Under the Americans with Disabilities Act, the individual is regarded as having a disability if he or she has

- a physical or mental impairment that substantially limits one or more major life activities,

- a record of such impairments, or

- been regarded as having such impairment.

Some have criticized the process of labeling a child as disabled in a particular way on the grounds that it tends to stigmatize the child. In some cases, particularly with children who are mentally retarded or emotionally disturbed, this is likely to be true.

Others, taking a pragmatic view, argue that knowing the cause of the disability is merely one step toward intelligent treatment and the delivery of relevant assistance. They argue that there is nothing inflammatory about calling something what it is. For example, everyone would agree that blindness is a condition that has certain implications for instruction. Accordingly, referring to a child as blind is more informative than simply saying that she has special needs, for eventually those needs must be identified before anything meaningful can happen in the classroom.

Terminology in the area of special education is constantly evolving. Although the term *handicapped child* was officially and correctly used for many years, it has been replaced by the term *child with a disability*. By now the offensiveness associated with the use of inappropriate language such as "cripple" and "midget" should be clear to all. Teachers should take the time to discuss appropriate terminology with the special education teacher. Above all, they need to understand that a functional limitation in one area does not mean a child is helpless in that or other areas or is unable to benefit from a wide range of experiences.[3]

FUNCTIONAL DISABILITY AND ACCOMMODATION

Current views of child development recognize that functional limitations are not absolute, but can, in many cases, be accommodated in one way or another. For example, it no longer is true that blindness automatically precludes learning to read ink print materials. For more than twenty years, technology has been available that enables these students to read the texts used in the class. Recently, technological advances have greatly speeded and facilitated the use of these reading devices. Nevertheless, their cost has kept many visually impaired children in the K–12 grades from gaining access to the equipment. Consequently, Braille is still used for most reading by blind students.

More important, teachers should realize that blind college students in a science course of study and blind adults employed in the sciences are much more likely to have access to new technology and to use it frequently. This means that regular teachers in the primary, middle, and secondary grades should avoid making speculative judgments about the capabilities of children with disabilities as the children become adults. It especially means that teachers should not preclude students from being involved in relevant activities on the basis of their own restrictive, and possibly erroneous, assumptions.

THE RELATIONSHIP BETWEEN DISABILITIES AND BARRIERS

People with disabilities typically encounter three types of barriers: attitudinal, personal, and environmental.

Attitudinal barriers

In the view of many people with disabilities, attitudinal barriers are the most frustrating of all. They occur because of misunderstanding, ignorance, and even bigotry. Attitudinal barriers are typically created by people without disabilities who have little or no experience with people with disabilities and, lacking evidence to support their view, reject disabled persons, ignore them, or otherwise preclude them from opportunities to reach their potential. Attitudinal barriers should not exist in schools and should certainly not exist in science classes, where premature judgments and faulty conclusions are actively discouraged.

Attitudinal barriers can be extremely frustrating. They are especially inappropriate in science classes.

Personal barriers

Personal barriers are associated with the disability itself and recur in many different situations. Clearly, a loss of vision makes it hard to acquire information visually, whether it is in print form, through a microscope, on the chalkboard or computer screen, or through film, television, or other visual medium. Similarly, the inability to hear interferes with the give and take of verbal communication. The inability to use one's hands, arms, or legs acts as a barrier to certain kinds of activity. Sometimes these personal barriers can be eased by providing the individual with a prosthetic device that facilitates the activity or with a helper who can augment their skills.

Personal barriers are limitations arising from the disability. Students with disabilities learn to cope with these limitations as much as possible.

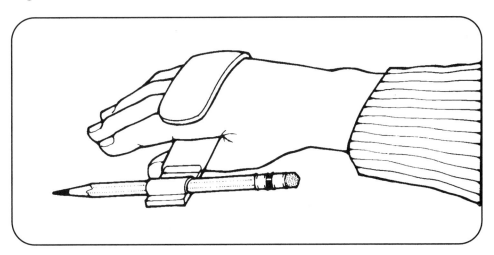

Environmental barriers

Environmental barriers occur in the settings in which the student must function. They include things like lab tables that won't accommodate a wheelchair, inaccessible supply rooms, and hazards carelessly left in the way of persons with low vision or blindness. Many environmental barriers exist because people without disabilities failed to design facilities with disabilities in mind. Modifying the environment to accommodate one person with a disability may seem expensive, but in the context of education it really means reducing that barrier for a number of

Removing or reducing environmental barriers is often a one-time task that benefits many students with disabilities.

students who may have a similar problem. Over the long run, a single solution can benefit many individuals.

As a teacher, you are already committed to the broad goals of education. Now we are asking you to make a special commitment to students who, because of some kind of disability, are faced with barriers and problems when they encounter "standard" science methodology and teaching.

Expand your ideas about the curriculum and your teaching style. You'll be pleased with the resulting positive impact on students with disabilities.

We are asking you to expand your ideas about the curriculum, your teaching style, and your knowledge of students' learning styles to include and encourage these young people. You will be delighted with the results of your efforts and the effect you have on students who need your guidance.

USING THE BALANCE OF *SCIENCE SUCCESS* TO ADVANTAGE

The balance of *Science Success* is structured to give easy access to information used with different disabilities in various instructional contexts.

Here are suggestions on how to use the balance of this guide to advantage.

- Turn to Chapter 2 to explore the instructional management issues that arise when students with disabilities participate in the classroom, the laboratory, and in the expanded classroom (outside of school).

- Turn to Chapter 3 when you seek ideas for avoiding and minimizing the effect of attitudinal, personal, or environmental barriers.

- Turn to Chapter 4 when you are preparing an instructional activity and you want to plan for ways to successfully engage the student with a disability. This chapter provides numerous examples of alternative strategies specifically addressed to different disabilities and taking place in different settings.

- Turn to Chapter 5 when you want to know how to supplement the resources in your school and where to get further information.

- Finally, turn to the Appendix if you are in a position to conduct an inservice workshop or to teach a preservice science education class. By sharing these materials, the entire staff can act in a coordinated, mutually supportive way to help students with disabilities reach their potential in science classes.

CHAPTER I NOTES

1. Teachers who would like to get in touch with persons with disabilities who are local and hold scientific and engineering positions should obtain a copy of the *Resource Directory of Scientists and Engineers with Disabilities,* edited by Virginia Stern, Diane Lifton, and Shirley Malcolm, and published by the American Association for the Advancement of Science, in Washington, D.C. The professionals listed have agreed to be resource consultants and most have had experience as mentors to young students with disabilities.

   Teachers who want to provide information to secondary students with disabilities about college and career planning in science, mathematics, and engineering are referred to two guides published jointly by the American Institutes for Research (AIR) and the American Association for the Advancement of Science: *You're in Charge* by Virginia Stern and Phyllis DuBois, and *Find Your Future* by Phyllis DuBois and Robert Weisgerber. The guides challenge students with disabilities to think through their career options and provide practical advice from mentors with disabilities who are active in the sciences. Inquiries about the availability of these booklets should be addressed to the American Association for the Advancement of Science, 1333 H. St. N.W., Washington, DC, 20005, Attn: Director, Project on Science, Technology, and Disability.

   Teachers who are interested in reading in-depth profiles about college students in the sciences and working scientists, engineers, technologists, and medical professionals with disabilities are referred to *The Challenged Scientists: Disabilities and the Triumph of Excellence* by Robert Weisgerber, published in 1991 by Praeger Publishers, New York, N.Y. This book reports on a study to identify the critical factors that influence success in entering and advancing in scientific fields.

2. A detailed analysis of the Americans with Disabilities legislation can be found in *The Milbank Quarterly,* Volume 69, Supplements 1 and 2, 1991. *The Milbank Quarterly* is available from Cambridge University Press, Journals Dept., 110 Midland Ave., Port Chester, N.Y. 10573.

3. Although certain disabilities do limit physical functioning, it is a mistake to underestimate what persons with disabilities can undertake. The recent book *Go for It! A Book on Sport and Recreation for Persons with Disabilities* illustrates clearly and impressively the wide range of activities engaged in for fun and exercise. Published in 1989 by Harcourt Brace Jovanovich, this book can be a powerful motivator for young persons with disabilities.

Planning Instruction for Students with Disabilities

T here are many ways to teach science. The ultimate criterion of whether the strategy a teacher uses is "good" or not is whether the students learn about science and can apply information in productive ways. A second criterion is whether the students develop personally, being able to think for themselves and to make informed, thoughtful decisions about issues that affect them in science and life.

Students need to learn science and need to develop personally.

Whether the students are just being exposed to science at the earliest stages, such as in the pre-reading grades, or are exploring complex subject matter at the secondary level, they need to acquire new information. Moreover, they need to learn how to acquire information in a way that encourages the following five elements:

- objective inquiry
- gathering of data

- understanding of concepts
- development of principles
- rational decision making

Given appropriate support, these students can acquire science information in a way that encourages use of scientific methods.

Intellectually able students with disabilities can participate meaningfully in science classes when the focus of instruction is on these elements. They need to have the chance and they need appropriate support.

Without attempting to "protect" the student with a disability, check regularly and informally on whether he or she is doing OK or whether something that has taken place in the class, such as a discussion point, assignment, or demonstration, needs to be repeated or reviewed. As will become clear, many of the regular activities of the classroom can be facilitated with the help of a buddy or team member.

DISABILITY-SPECIFIC GUIDELINES

You can take steps to make information more accessible to students with disabilities.

Whether teaching in the classroom, the science center, or in the field, be aware of whether what you show or say can be seen or heard by the student with a disability. Take appropriate steps to see that the information you are trying to convey is accessible to the student. Here are some of the steps you can take for six different disabilities.

Guidelines for students who are blind

Arrange to have materials converted to Braille. Allow ample lead time.

- Arrange in advance with the special education teacher to have handout or text materials converted to Braille for students who are blind. Both typed and written materials can be prepared in Braille. Lead time will vary—it is likely to take a week or more depending on the volume of material.

Arrange for a reader (paid or volunteer) when the situation warrants it.

- Alternatively, when substantial amounts of material must be covered in a short period of time, ask the special education teacher to arrange for a reader for the student. Also use an aide when film or television materials are being shown to provide the student with simultaneous commentary about the visual action.

Convert pictorial and diagram materials to a raised-line format.

- When pictorial or diagram materials are to be displayed, they can often be converted to raised-line diagrams using Thermoform materials. If diagrams are too complex, try using a three-dimensional model instead. Ask the special education teacher about these alternatives.

Repeat aloud what is being written on the board.

- When information is written on the board, or when a transparency or slide is used, repeat aloud what is being written on the board and orally describe what is on the screen. The same step is appropriate for demonstrations.

Be sure test materials are in a "fair" format and administered in a fair way.

- When tests are to be given, be sure items are prepared in an appropriate format so that the assessment accurately reflects the student's knowledge of the subject matter rather than his or her ability to interact with the test.

Guidelines for students with low vision

Determine whether a video enlarger or large-print materials are available for the student's use.

- Discuss with the special education teacher whether an electronic video enlarger can be obtained so the student can read regular-size texts and handouts. If not, explore the possibility of getting large-print versions of print materials.

- In many classroom situations, such as when holding up an object or chart, pointing out something on the board, carrying out a demonstration, or examining a science specimen, be sure the student with low vision can see.

  Make sure the student is positioned for maximum visibility in the classroom.

- Some students with low vision can only see an object when holding it very close or off to one side. Sometimes they will only be able to see one part of a large object at a time, so allow extra time for the student to assimilate the information.

  Allow extra time.

- Pay particular attention to classroom lighting—darkness, low-contrast, and high-glare situations (such as window glare) should be avoided.

  Pay attention to lighting and glare.

- For both blind and low-vision students, references to distant or untouchable objects are like references to an abstraction (e.g., an unreachable object such as the moon or a bird in a tree). Naturally, scientific principles that become clear to sighted students in such situations may have to be restated.

  Be aware that untouchable objects that are visible to other students will be abstractions to the student with low vision.

Guidelines for students who are deaf or hard of hearing

- The difference in degree of deafness will influence your strategy for teaching. It makes a difference whether the student has mild, moderate, severe, or profound hearing loss. Discuss with the special education teacher whether there is a need for an interpreter (perhaps a community volunteer). The student may have been taught to read lips or to read sign language—either (a) American Sign Language (ASL), which has a very different syntax from English; (b) Signing Exact English, which approximates English syntax; or (c) total communication, which includes lip-reading and signing. These are entirely different strategies for communication, not unlike different languages. If you do not use an interpreter, speak very clearly and at a reasonable pace. Make sure the student is positioned in such as way as to pick up as much as possible of what is said by lip-reading, and supplement speech with written notes.

  Determine the student's customary mode of communication, such as sign language, and clarify the availability and use of interpreters.

- Because lip-reading will only partially transfer information (some sounds cannot be made visible), repeat key statements, and look for confirmation that the

  Repeat key statements, and look for confirmation.

student has understood. You should be aware that lip-reading is harder when the speaker has a thick mustache, no lipstick, or a pronounced accent.

Remember that lip-reading is not possible unless the student can see the speaker clearly.

Have someone make duplicate notes so the student can concentrate on the speaker.

Explore the possibility of sound amplification, controlled by the student.

Try to locate captioned media for use in class.

For students who have difficulty speaking clearly, be encouraging and develop patience and support in the student's classmates.

Be sure the testing procedures and tests are fairly administered to the deaf student.

- Lip-reading is impossible if the speaker is turned away or the room is darkened. Distance and location in the room are also important. Try not to stand too far from or too close to the student. If possible, supplement what is said with written, print, or graphic materials.

- For students who lip-read, arrange for someone in the class to make duplicate notes (using carbon paper for the duplicate or a photocopy machine, if available) so the individual can concentrate on the speaker instead of having to look back and forth.

- For hard-of-hearing students, ask the special educator about the possibility of sound amplification, using wireless transmitter/receiver equipment that lets the student control the volume of speech received.

- With the special education teacher, explore the possibility of using captioned films or television. Allow plenty of lead time. If these are not available, provide a review of the main points of the presentations to the student in another form.

- In class recitations, many deaf and hard-of-hearing students have difficulty speaking clearly (students who have become deaf later in their lives may be an exception). Encourage the student to make the effort but also alert the other students that they will be expected to listen very carefully. In practice, hearing students who have grown up with deaf students will have relatively little difficulty in understanding their speech.

- When testing deaf students who depend on ASL, keep in mind that English grammar and usage is very different from ASL grammar and usage. As a consequence, deaf students may need more testing time and may have greater difficulty in expressing themselves orally or in writing. This difficulty can mask students' knowledge of science.

- Students who cannot hear may have difficulty in speech production as well. If the student is expected to give a report, respond in class or converse with others in a team activity, take steps to see that the experience is productive both for him or her and for the other students. Provide a supportive, patient environment. Try to identify several students who can act as quasi-interpreters. If a presentation is substantial, and the student needs a regular interpreter to take the pressure off and be clearly understood, make the arrangement beforehand with the special education teacher.

  Consider the student's possible difficulties in producing speech.

- Monitor progress frequently to ensure that students who are deaf or hard-of-hearing are receiving information.

  Monitor progress.

Guidelines for students with physical disabilities

- Because there are many different kinds of physical disabilities, have a meeting with the special education teacher to go over any medically related limitations on activities that the student might have. For example, with a severe disability such as spina bifida, it might be important for the student to change from a sitting to a reclining posture after a period of time.

  Become informed about any medical limitations that might affect the student's activities.

- Some students with physical disabilities will tire more easily than others, which can effect the type of assignments they can be given. Others may need to leave the room for health care reasons. Flexibility is important for these students.

  Consider the effects of fatigue and health care needs on the student.

- For students in wheelchairs, locate them where they can have access to hands-on activities, see demonstrations, and move around the room to the maximum extent possible.

  Be sure students in wheelchairs can participate in activities as fully as possible.

- For students with little or no hand use, have a student "buddy" obtain materials that are not easily reached. Have the student with a disability "direct" the buddy in the task that he or she is unable to do, thereby demonstrating a knowledge of how it should be done.

  Assign a buddy who can help to reach or manipulate objects as needed.

- When off-site field trips are being planned, anticipate special problems that may be encountered for the student with the physical disability, and try to work out an accommodation in advance with the host organization.

  Anticipate problems that might arise, and take steps to minimize negative impact.

- Students with health impairments present a special case within the group of physically impaired. In addition to some physical limitations, they may have attendance irregularities that make it hard for them to keep up. Consult with the special education teacher on ways you can keep the student involved through manageable outside assignments. As much as possible, try to include home-oriented tasks that parallel hands-on, classroom activities in this outside study.[1]

  Have a contingency plan to keep students with health impairments involved even though they may have more frequent gaps in attendance.

Guidelines for students with learning disabilities

- Understand the specific nature of the learning disability for each student. Modify your instructional adaptations to reflect whether the disability is in reading, mathematics, or some other area.

  To make appropriate adaptations, understand the specific type of disability.

- Present information in a clear, well-organized manner. Use straightforward instructions, with step-by-step unambiguous terms. Check on the student's understanding of assignments.

  Avoid ambiguity in instructions and in giving information.

- Help the student build confidence in his or her oral communication. Begin with short-answer responses by asking questions that can be answered yes or no, then gradually increase expectations as the student gains confidence.

  Ask questions in a way that helps the student gain confidence.

- Use multimodal materials to involve the different modalities of learning. Balance textbook and hands-on instruction, and use graphic and audiovisual materials if appropriate.

Allow extra time for tests.

- Allow extra time for completing tests, or use oral tests. Consider cross-age or peer tutoring if the student appears unable to keep up with the class pace.

Guidelines for students with speech impairments

Focus on what is said, not how well it is said.

- Keep in mind that speech impairments can easily mask the extent of knowledge a person has about a subject. Accordingly, don't focus on how the student speaks but upon what is being said. Often, other students in the room will be able to understand the speech-impaired student fairly well, so the focus of the problem may be your own perception and understanding. Patience and practice will likely take care of any communication gap.

Let the student know you welcome his or her participation.

- Make sure the student knows that his or her class participation will be welcomed, and ask the student privately how he or she would like to handle situations such as responding to questions in class.

Use an interpreter if necessary.

- Severe speech impairment may require the use of an interpreter upon occasion, especially at the beginning of a school year in a new environment. Consult with the special education teacher about the type of interpreter that would be appropriate given the expressive limitations of the student and whether he or she knows sign language.

Investigate specialized technology to supplant vocalization for some nonvocal students.

- Specialized computer technology (hardware and software) and special purpose communication boards can be used effectively in lieu of speech by students who speak only with difficulty or who cannot speak. (Communication boards are any of a variety of devices or display boards that allow an individual to choose letters, words, or symbols that convey meaning analogous to speech.) Consult with school district staff in the area of computer technology to identify appropriate equipment.

Guidelines for students with emotional disorders

- Consult with the special education teacher about interaction techniques that have been used successfully with a particular student. Try to establish whether the student's emotional state is sensitive to particular classroom situations, such as working in groups or making a presentation. If so, and if you expect that these situations will arise, involve the special education teacher in mapping out a strategy to minimize the potential problem.

- Find out whether the student is on medication, and ask about its maintenance schedule and side effects. If the student is taking medication or is irregular with its use, his or her classroom performance can be negatively affected. In such case, keep in mind that the difficulty does not reflect on the student's ability in science, but is a result of the medication.

- Be sensitive to team pairings in cooperative-learning situations, and have a contingency plan. Try to assign teams that will support rather than alienate the student.

Learn about techniques for interaction that have been effective with the student in the past. Anticipate classroom situations where the student's emotional state will be vulnerable.

Find out whether the student is on medication, what the schedule is, and what the effects may be on his or her performance in class.

Be sensitive when making team pairings so that the student is supported.

COLLABORATION WITH THE SPECIAL EDUCATION TEACHER

Special education teachers have special training that prepares them for working effectively with students with disabilities. They are a resource you will want to call upon early and often as you work with a student with disabilities. Below is a summary of some of the specialized skills that special educators have. By collaborating with your school's special education teacher, you will be expanding your own teaching capabilities.

Feel free to involve the special education teacher without hesitation as

- a partner in advance planning

- a source of information about techniques

- a source of information about the student's needs and abilities

The special education teacher knows the specific needs of a student with a disability and can support you in many ways.

The special education teacher can collaborate in a variety of ways.

- a possible source of adaptive equipment or ideas
- a source of help with adapted or specialized instructional materials
- a resource for obtaining help from other agencies

TEACHING IN THE EXPANDED CLASSROOM

To this point, attention has been given to teaching practice in regular science classes. However, as you have probably found, good science often takes place outside the regular classroom—in science centers, laboratories, on field trips, and on the school grounds. Science projects that are undertaken by teams of students are likely to require some information gathering outside the classroom, and many science-related activities are best made clear by showing how they work in natural settings.

For example, while growing plants from seeds is done frequently in classrooms, it is more compelling and memorable if the students can share the challenge of growing seeds outdoors. Moreover, as students plant some seeds outdoors and some indoors, the idea of contrasting growth environments can be introduced—the classroom may be more favorable (e.g., climate controlled) or less favorable (e.g., less sunlight).

Students appreciate the opportunity to think in terms of what happens outside of school and how science is applied in nonschool settings. They become more aware of the generalizability and interchangeability of science information gained in the school setting and in other community settings. In planning these out-of-school experiences, consider choosing locations where a role model with a disability will be found, such as a field trip to a science-oriented business where a scientist with a disability works together with nondisabled employees.

There may be some initial concern over involving a student with disabilities in these "expanded school" settings. This concern should not lead to a rejection of

Good science teaching takes place in a variety of locations both in and out of school.

Experiments can be developed to contrast the different environments in which learning takes place.

Involve students with disabilities in learning experiences in expanded instructional settings in the community.

Involve students in expanded experiences to prepare them for independence as adults.

the experience for the class or, especially, for the student with a disability. If the student cannot go to the outdoor location because of asthma or some other problem, perhaps some part of the study area can be brought back for the student to examine.

In addition to conveying knowledge and skills to students, educators are committed to preparing them to becoming independent and self-sufficient as they enter adulthood. Hands-on experience can be a sort of pre-vocational experience, which can benefit all students. No one in the class needs this kind of practice more than the student with a disability, and none is more likely to appreciate the opportunity to participate.

SPECIAL SITUATIONS IN COOPERATIVE LEARNING

Cooperative learning implies that students are partners in learning. They take an active rather than passive role, and they typically share the tasks involved. This means that learning becomes focused on how to ask "scientific" questions and discover answers using "scientific" processes. It is much less focused on how much information can be recalled from text or lecture sources. Cooperative learning is often seen by students as relevant learning.[2]

> Students see cooperative learning as relevant learning because it lets them work together to discover answers to questions.

In the course of a cooperative-learning lesson, students learn to complement each other's skills and to share information gathered by different people in different locations. In effect, cooperative learning is an approximation of how science occurs in adult settings, where no one knows everything and the pooling of resources, information, and talents is commonplace.

> Cooperative learning is similar to science at the adult level, where people work as colleagues, forming teams to use their complementary skills.

Just as is true in the real world, students engaged in cooperative learning should understand that the essence of being a team player is that the same (team) score is shared by all and that the score is a direct result of each team member making a contribution. In the real world, not all scientists are equally well versed in all aspects of the science they pursue. They turn to specialists when they need assistance, and they are willing to share their own expertise when called upon by others. They form productive teams. This is the model that should be in place in teaching science.

This model suggests that cooperative learning will work best when the teacher

- organizes teams with an eye to including a range of ability levels

- assures that members of the team can contribute different skills

> Students learn to be team players.

- makes each of the members of the team aware that the team as a whole will be evaluated and not just a single member of the team, thereby helping students recognize the unique contributions that each can make to the benefit of all

The student with disabilities should be a team player also. The question is, in which team? In answering this, consider this piece of research that has emerged from experimental, comparative studies of cooperative learning in school settings. There appears to be no *significant difference* in team performance between teams that are composed of students with varied abilities and those that are more homogeneously organized according to ability level. However, there *is* a difference in the socialization that takes place within the two types of teams, and that difference favors teams that have mixed ability levels.[3]

> Research has shown that participation by students with sharply different abilities does not impede team effectiveness.

Clearly, the team member who has a disability should be able to find a productive niche. If care is taken in the introduction of cooperative learning and attention is paid to how well the teams are functioning, the student with a disability will get a big boost in self esteem.

USING ADAPTIVE TECHNOLOGY

Students with disabilities can often benefit from using some type of technology. Technology can be thought about in these ways:

- Technology can consist of some type of personal device, tailored to the individual, that enables the student to perform an activity that otherwise would not be possible or readily accomplished.

- Technology can be designed in a generic way to affect the environment and benefit any number of people with disabilities who use that environment.

Among the many mechanical and electrical devices used are a wide range of mobility devices.

Visually impaired and blind individuals have become increasingly independent as travelers, in part because of the availability of electronic devices, such as the laser cane, that can help them to sense and navigate their environment. While some of these devices may not be widely used by students with disabilities, their

availability helps to open the range of job opportunities these students will be able to consider as adults.

Wheelchairs for the physically disabled come in many different configurations, ranging from sport chairs to all-terrain chairs, chairs that can climb stairs, chairs that elevate to an upright or standing position, electric chairs that can be controlled merely by toggling a joystick, puffing through a tube, or even by tilting one's head to trigger infrared sensors. Wheelchairs are available in customized sizes and with a variety of padding and support.

Specially adapted sporting equipment, such as hand-propelled bikes and sit-skis, are no longer a rarity. The availability of specially adapted vans and cars allows many young people with disabilities to prepare for a productive future that involves driving themselves to work.

While the typical student will not have access to or need a wide range of mobility equipment, options do exist for severely physically impaired individuals. Rehabilitation engineers can work wonders. In some communities, where rehabilitation engineers are not available or are overloaded, groups of engineers from business and industry have volunteered to build or adapt special equipment.

Technology also contributes substantially in the area of **augmentative communication devices.** Students with severe speech disorders that impede intelligible speech can use one of many variations of an electronic communication board. Deaf students can use telecommunication devices to communicate by telephone. Among the types of technology that can be shared by several persons with disabilities are specialized reading devices. Benefiting students who are blind and severely visually impaired, and also certain learning-disabled students, are a number of devices that scan text materials and convert the information to spoken output. Other devices use closed-circuit television technology to allow user control over image magnification. This can facilitate general purpose reading or the examination of a range of small objects, from checks to labels.

A variety of communication devices are available.

Of course, **computers** represent a technology that has benefited many persons with disabilities. Sometimes computers must be modified slightly to make them user-friendly. Examples are specially designed keyboards that help guide the individual's hand toward the right keys; software that allows the computer to accept commands that are not dependent on multiple, concurrent key strokes; and computers that are voice driven.

Computers are a technology that can benefit all persons with disabilities.

Customized computer adaptations and special software are now available.

Computers have been a tremendous tool for people with severe disabilities, such as those having severe cerebral palsy together with an inability to speak. Artificial speech output from computers has now given some of these individuals a new capability. Through customized adaptations and special software, a number of cases have been documented where a person with a disability has shown extra-ordinary talent and problem-solving ability of which they might have been inca-pable before the computer opened up an avenue.

CHAPTER 2 NOTES

1. Warger, C. (1991) *Peer Tutoring: When Working Together Is Better Than Working Alone*. Research and resources in education, Research Report Number 30, December, 1991. This six-page research brief was prepared by the ERIC Clearinghouse on Handicapped and Gifted Children, part of the Council for Exceptional Children. It summarizes a variety of research and professional writing on the principal issues that arise in peer and cross-age tutoring.

2. Resources on cooperative learning include the following:

   Cohen, E. G. (1986) *Designing Groupwork: Strategies for the Heterogeneous Classroom*. New York: Teachers College Press.

   Slavin, R. E. (1991) *Student Team Learning: A Practical Guide to Cooperative Learning*. Third edition. Washington, DC: NEA Professional Library, National Education Association.

3. Hooper, S., and Hannafin, M. (1988) *Cooperative Learning at the Computer: Ability-based Strategies for Implementation*. In Proceedings of selected research papers presented at the annual meeting of the Association for Educational Communications and Technology, January 1988. In ERIC, ED 295647. Complete proceedings are in IR013331. This research report compared the achievement of high- and low-ability eighth grade students working cooperatively on computer-based instruction when they were grouped heterogeneously or homogeneously.

Chapter 3

Managing Instruction and Learning

In Chapter 1 we pointed out the three types of barriers encountered by students with disabilities:

- **Attitudinal barriers** involve interpersonal behaviors that adversely affect persons with disabilities, thereby limiting their access or opportunities.

- **Personal barriers** are associated with the disabling condition and directly limit the activities a particular individual can undertake.

- **Environmental barriers** negatively impact many persons with disabilities due to obstacles in the environment in which the individual studies, works, plays, and lives.

Students with disabilities encounter three types of barriers: attitudinal, personal, and environmental.

These barriers can influence science instruction, especially when it involves a high degree of dynamic interaction between the teacher and a class that includes students with disabilities. As a classroom teacher, perhaps you feel that you lack a detailed knowledge about different disabilities, have limited training in special education, or are inexperienced in meeting the needs of these students. This feeling is understandable. Fortunately, most adaptations and adjustments to accommodate a disability are easier, more straightforward, and more minor than you might expect. In the process of discovering these adaptations, you are likely to discover that the experience has expanded your own skill.

Classroom teachers may feel unprepared to meet the needs of students with disabilities, but many of the needed adaptations are minor.

Keep in mind that before they came to your class, students with disabilities have already encountered situations in which they had to do things differently from their nondisabled classmates. Some have become quite adept at coping, others less so. In either case, be aware that you'll be building on their prior experience and not starting from scratch.

These students bring prior coping experience to the classroom.

You'll quickly realize that the student with a disability is not an adversary but a welcome addition to the class. The student with a disability wants to be treated as normally as possible and responds warmly when given fair treatment. He or she is central to developing understanding and a sense of shared commitment in the class.

The student with a disability quickly becomes a welcome addition to the class.

BUILDING POSITIVE ATTITUDES AND RELATIONSHIPS

Teachers who have not previously had students with disabilities in their classes are often surprised to discover how satisfying, both professionally and personally, it can be to work with these students.

Examine your attitudes about teaching students with disabilities

Here are a few fundamental principles that will be important to you as you begin to interact with students who have disabilities.

Be positive and supportive.

- The student with a disability will reflect your optimism as well as your doubts, so emphasize the positive. Taken one step at a time, there is every reason to believe things will work out favorably.

Be reasonable in your expectations.

- The student with a disability may take longer in completing an assignment or test simply because there are variables present that prevent him or her from working as rapidly. This should not reflect negatively on his or her ability or understanding.

Recognize individual differences for what they are—not everyone has the same interests and aspirations.

- As students advance in school, some will move naturally toward science as a field of study and others will choose other career paths. The same holds for students with disabilities. Be careful not to extinguish the aspirations of students with disabilities if they show an interest in science. There *are* plenty of ways in which they will be able pursue their dreams, however difficult it may seem.

Examine the student's interpersonal skills and self-esteem

All students, whether they have a disability or not, must learn to get along with others and need self-esteem. Students vary considerably in their self-regard and interpersonal competency. Some are confident and take easily to learning and using social skills, while others seem to lag behind or avoid interaction.

Self-esteem and interpersonal skill are essential for all students.

Following are some of the reasons students with disabilities may have difficulty with social skills, along with suggestions about what you can do to help.

Check on the student's basic capacity to communicate.

- Consider the student's basic communication skills. Keep in mind that there is an important difference between the student who is reluctant to communicate but can, and the student who wants to communicate but has an impaired capacity to do so. Some students with disabilities have difficulty in communicating—being understood or understanding others—and feel cut off from the other students.

  While you cannot be expected to fill the role of a speech, hearing, or language therapist, you can encourage students to interact as best they can, show patience in hearing them out and in speaking to them, and informally encourage their classmates to be responsive when they try to engage in conversation.

- Check to see if the student's approach to social relationships is constructive. Some students with disabilities are overly shy and withdrawn, very sensitive, or otherwise actively discourage others who might approach them. It is helpful to know whether the student's own behavior is contributing to his or her isolation. Sometimes the student will act out or be defensive, or misjudge classmates' willingness to interact.

Examine the student's social relationships to see if they are constructive and encourage others to be friendly. Implement a buddy system.

The student's perceptions of how others feel about him or her may be accurate or inaccurate, but the importance of helping the student develop social skills cannot be overemphasized. Consider implementing a buddy system in which the student will have an opportunity to develop a friendship with a caring student.

- Check on the student's self-esteem. Some students are more confident, adjusted, and self-accepting than others. There is likely to be a wide range in any class. Students who have little confidence may feel that way much of the time or may only feel that way when they are in a situation in which they are uncomfortable.

Examine the student's self-esteem. Help the student to feel as though he or she has something worthwhile to contribute.

Make the student feel that he or she *belongs* in the science class and has something to contribute. The key is to actively involve other students in helping (and praising when appropriate) the student with a disability.

- Check on the student's assessment of his or her science potential. Science can be a wonderful career area for intellectually able students with disabilities. However, science can also be intimidating to students who have no expectations of becoming scientists as adults.

Try to recognize whether the student has a natural interest in science. Encourage students to learn science so they can apply it as adults.

Through the kinds of educational experiences you provide, help these latter students to understand that even if they don't have a natural flair for science, a knowledge and application of science processes will still be valuable to them as adults.

Develop the student's self-confidence and willingness to participate

Students with disabilities are especially challenged when they are expected to interact with nondisabled students in regular classes.

Any student with a disability who has been placed in the competitive environment of a regular classroom is especially challenged. If the student has had little prior experience in regular classes, or the prior experience has not been altogether satisfying, he or she can feel intimidated and even afraid.

Here are a few steps to follow in helping the student develop self-confidence and overcome any reluctance to participate.

Strike a balance between asking too much or too little.

- Bring the student along at a pace that makes success likely. Try to strike a balance between asking too much or too little of the student with a disability. For the most part, these students want to be treated as their classmates are; however, it is a lot better to set objectives that can be reached than to "dump" excessive work on the student from the beginning. Involve the student in planning work loads. Be sensitive to the fact that it will take longer for some students with disabilities to do homework.

Acknowledge the contributions of the student with a disability.

- Recognize the contributions and achievements of the student. Even though the progress may seem slight, to the student with a disability it may be a significant accomplishment. In any case, your recognition lets him or her know that you are aware of his or her efforts. As you compliment the individual, don't be too effusive, but let the student know why you are proud of the way he or she is progressing.

Provide opportunities in which leadership can emerge.

- Provide opportunities for the student to exhibit leadership. If the student is adjusting well to working with his or her classmates, provide opportunities in which leadership potential can emerge. Have the student play an important part in designing activities, organizing study teams, reporting on the outcomes of science activities, and otherwise demonstrating that he or she is more than just a follower, not someone who has to be helped at every turn.

HELPING THE STUDENT DEAL WITH PERSONAL BARRIERS

Each disability presents certain barriers of a personal nature. Some may be eased through technology, but even that is limited in terms of what can be expected.

Each disability brings with it a set of limits on an individual's performance. These limits can be stretched by personal determination and commitment and by the introduction of technology that enables the individual to do something that otherwise would be virtually impossible. Regular and special education teachers are instrumental in the former case, while instructional technologists, engineers, and rehabilitation specialists are important in the latter case. The task is straightforward, but not necessarily simple.

Here are some strategies for helping the student with a disability stretch his or her limits.

Encourage the student to try practical, safe solutions to personal problems.

- Encourage the student to try unusual—but safe—solutions to everyday problems. Regimented thinking about "the right way" to do a task can be counterproductive if that way is not the best way for the person with a disability.

To build independence and self-reliance, involve the student in thinking about solutions to problems that arise. What may seem awkward for one person may be practical and effective for another. Safe trial-and-error is an appropriate principle to follow in trying alternative procedures. This helps the student to develop a sense of control over outcomes and to increase his or her skill in decision making. It can also give the student experience in experimentation, a key method of science.

- Identify an appropriate mentor or role model. Try to pair the student with an adult with a similar disability, particularly one involved in science, who can serve as a role model.

  Try to put the student in touch with an appropriate role model.

  If that is not possible, expose the student to information about role models through print or other media. Challenge the student to follow a career path pioneered by others with disabilities and to pioneer on his or her own.[1]

- Obtain information from community organizations that have specialists who are knowledgeable about a particular disability. Because community specialists generally encounter a wide range of problems, they may be aware of new developments or solutions.

  Contact local community organizations who serve people with disabilities to find out ways they can complement your efforts.

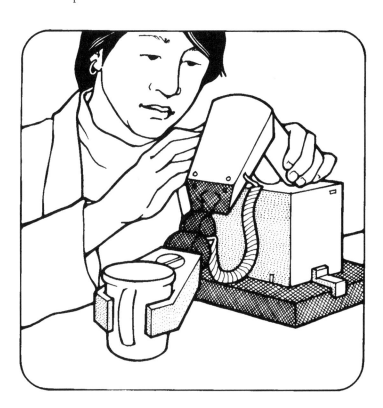

For example, an agency serving the blind should be up-to-date in its knowledge about the costs and availability of assistive technology. Similarly, rehabilitation engineering centers are at the cutting edge of new applications of technology in *prosthetics* (substitution or replacement for missing parts) and *orthotics* (support or bracing of weak parts). As an example, functional electrical stimulation (FES) is making it possible for some paralyzed individuals to regain partial functioning of limbs, and this can dramatically affect their independence. Unfortunately, engineers communicate with the school staff about these latest developments only sporadically, if at all.

- Obtain information from instructional technologists in the school. They can be an important source of information about new software and hardware in the marketplace that can make the difference for a person with a disability. Assistive technology can include both customized tools, switches, and controls as well as general devices that make complex tasks easier or quicker. Computer technology is one area in which the state of the art is continuously being

  Consult with educational technologists in the school district to learn more about adaptive software and hardware.

MANAGING INSTRUCTION AND LEARNING

27

advanced, allowing persons with disabilities to do things they could not do before. A similar situation exists for closed-circuit, portable image magnification and various telecommunications devices.

MANAGING ENVIRONMENTAL AND INSTRUCTIONAL BARRIERS

Science instruction and learning takes place in many different environments: regular classrooms, science centers or laboratories, and in the field in community settings, including the home. Each environment can present special educational challenges, and special problems for one kind of disability but not another. Chapter 4 provides strategies and case study examples of various ways in which environmental and instructional barriers can be overcome.

Science learning takes place in multiple environments that present different kinds of challenges for effective teaching.

Look at the Framework for Instruction chart. It is organized into a pattern for effective instruction, which is explained further below.

There are four key elements to effective teaching in science.

A. Planning activities, before instruction

B. Effectively managing instructional activities

C. Encouraging students to use science processes as they "do" science

D. Assessing student development

Framework for Instruction			
A. Plan in Advance	*B. Manage Instruction*	*C. Use Science Processes*	*D. Assess Student Growth*
Resource planning	Teacher's activities	Design of projects	Fair testing practices
Coordination	Student's activities	Manipulation of objects	Performance assessment
Advance planning	Cooperative learning	Observation	Written testing
	Routine activities	Data collection	Oral reporting
		Interpretation	Portfolio assessment
		Concept formation	

Plan in advance

Try to anticipate instructional barriers and problems, and take appropriate action beforehand. It is not necessary to plan every last detail, but it is important to think ahead about situations that are likely to arise and, early on, take steps to facilitate the whole process.

Identify, develop, and organize human resources in the school, the home, and the community.

• Identify, develop, and organize human resources. Find out who can help and how. The special education teacher should be in a position to provide a substantial amount of support, whether it is in preparing adapted materials (such as brailled materials), identifying outside resource people, locating adaptive equipment owned by the district, or helping the student settle in to the routine of science instruction in the regular classroom.

Part-time classroom aides, specialized staff (e.g., interpreters), and tutors may be available, as well as peer or cross-age tutoring programs using students from higher grades. Without advance planning and justification, these are unlikely to be available, however. Community-based resources (such as people with disabilities similar to those of the students in your school) and parent volunteers can be invaluable sources of ideas and support. The earlier you learn about them, the more effectively you can use their help.

- Identify, develop, and tap appropriate physical resources. Science lessons often involve materials and apparatus that must be assembled and used. The availability of these items may be limited by school budgets, and the items themselves are usually standard in design and use. Take stock of the science equipment and supplies in the school. If the student with a disability is going to use these, give advance thought to how he or she will do each of the tasks.

  Identify, develop, and organize physical resources.

  For example, premeasuring materials may facilitate progress without changing the basic nature of the experience. Similarly, advance planning might call for substituting unbreakable chemically nonreactive containers for breakable ones.

- Coordinate across resources. Home-school coordination can help the student by making sure that what is accomplished in one setting is used in the other. For example, if a procedure or technique has been worked out at school that facilitates the manipulation of objects (or some other activity), perhaps it could also be applied at home. Similarly, coordination between special education and regular education staff is essential. Instructions, textual materials, and tests can be brailled but they will only be useful if the content is specified and the brailling is completed with enough lead time so that they are available when they are needed.

  Coordinate across human and physical resources: home and school, within the school, and between the school and the community.

  Instructional staff in and outside the school can coordinate their efforts to great advantage. For example, in planning field trips, it is helpful to inform the staff at the host location (science museum, nature habitat, etc.) that a student with a particular disability will be a part of the group. This gives them time to take whatever steps they can to make the experience relevant and effective.

Manage instruction effectively and creatively

As the manager of instruction, you can employ a variety of instructional strategies that may vary depending on the setting (classroom, science center, or field setting) and the goal of instruction (e.g., develop knowledge, skills). In all cases, gear your choice of an instructional strategy to the capacities of the student with a disability. Here are several considerations that should enter into the choice of a strategy.

- Communicate directly with the student about ways in which the instruction/learning process can be facilitated. Make the student feel comfortable about telling you when something has not been presented in a way that was helpful.

- Be attentive to cues from the student as to whether he or she is actively engaged. As instruction proceeds, be aware of the student's involvement. Is the student taking notes and participating in classroom dialogue? Is the student able to understand and act on instructions? Does the student know what to do and how to do it? Is the student asking questions that reveal awareness, or does the student seem confused?

- Try new ideas and approaches. While there is merit to a structured pattern of instruction, a pattern that can focus the student's energies on the task at hand, there is also merit in varying the routine to provide a student with a fresh perspective. A mix is effective when it provides first-hand experience to complement science knowledge gained through lectures or reading. Similarly, peer tutoring or cross-age tutoring can be helpful.

- Have a contingency plan. Be flexible—if one approach doesn't work, be ready to try something different. Use your imagination and creativity. Try to make learning fun for all, including the student with a disability.

Focus on applying science processes

At all levels, from national commissions to district-level teachers on curriculum planning teams, there is agreement that our society depends upon an increase in the number of students who choose science as a career and, equally important, an informed electorate that can make judgments about issues based on a rational consideration of the facts. "We learn best when we learn by doing" is an oft-repeated phrase. It is certainly true that those students who think of science as something that other people (i.e., scientists) do will have a very different attitude toward science than those who have been directly involved in science activities. By giving students "hands-on" experience, teachers can build an appreciation of science methods as well as awareness of scientific facts.

Here are the key elements of involving students with disabilities in hands-on science.

- Involve the students in the design of experiments and projects. If students with disabilities know about what's ahead, they may have some ideas about how they can participate effectively.

- Involve the students in the manipulation of materials and objects to produce results. Though certain activities may be unsafe or impractical for particular students, it is nevertheless true that these are more likely to be the exception than the rule. By becoming involved in various demonstrations and experi-

---

Different strategies may be useful in the classroom, science center, and field settings, but the instructional strategy selected should be geared to the student's disability.

Communicate directly with the student during your planning to get his or her suggestions.

Pay close attention to the student during instruction to see that he or she is productively engaged.

Use a pattern of instruction to assure consistency, but vary instruction across modalities and strategies to keep it lively.

Have a contingency plan for when things don't work out as expected.

Instruction should focus on strategies that will produce students who are science literate and have a "hands-on" grasp of science processes.

Involve students in the design of experiments.

Involve students in the manipulation of materials and objects to produce results.

ments, students with disabilities will gain self-confidence as well as a better understanding of the underlying scientific principles.

- To whatever extent possible, emphasize the importance of careful observation. Observation of phenomena, processes, actions and reactions, and their relationship to timing and environmental factors are all important to the collection of accurate data. Effective observation involves knowing what to look for and how to recognize something significant when it occurs. For some students with impaired sensory functioning, observation may have to be aided by a partner.

  Stress the importance of careful observation. Allow for alternative methods of observation if necessary.

- Involve the students in recording and organizing information. The collection of data is only complete when it has been accurately recorded, organized, and synthesized. This process of examining data to establish a result or to draw a conclusion can be thought of as *descriptive analysis*—what the data showed.

  Involve students in recording the data they observe and displaying it in appropriate formats.

  For example, measurements of an experiment taken over time should be logged as raw data and, if appropriate, displayed in a graph or in another format to organize the information in a meaningful way.

- Involve the students in interpreting and generalizing findings. Interpretation is an important step in which students must be actively involved. They should be responsible for stating not only what they found out but also what it means. In making interpretations, they should be advised to stay close to the evidence from their previous findings; that is, not to inject unsupported speculations and claims in making their interpretation.

  Involve the students in interpreting and generalizing from the data.

  At the same time, they should be encouraged to think broadly about possible implications of their interpretation in different situations. Critical thinking involves knowing when and how to generalize findings to specific, new situations to make informed decisions about what is likely to happen.

- Guide the students in forming concepts about science and science methodology that will be relevant in their lives. Gaining insights into underlying scientific relationships prepares students for effective problem solving in their lives. It is essential that students gain a feeling for the many interrelationships that occur in nature and how these relationships can be influenced by things that we do as individuals and as a society.

  Guide students in forming concepts that will be relevant to their lives.

Assess growth and development with flexibility and fairness

Assessment is part and parcel of science education. Under the law, reasonable accommodation for persons with disabilities applies to assessment as much as to instruction. The teacher's responsibility is to give the student with a disability a fair opportunity to compete with students who have a clear advantage merely because they don't have to contend with the disability.

Fairness in testing and assessment can be accomplished through reasonable accommodation.

Formal, standardized assessment is used to establish the achievement of students in various subjects and, taken collectively, gives us a "reading" on how well American students are doing in comparison to students from other countries.

At the classroom level, teacher-developed assessments are used to evaluate the accomplishment of milestones (e.g., end-of-unit tests), as benchmarks of student progress (e.g., midterm examinations), and as a basis for grading (relative performance of students against a standard). Students with disabilities should be included in these assessments provided that reasonable accommodation has

Students with disabilities should be included in regular classroom assessment to the extent that it is feasible to do so.

taken place and the intellectual demands of the examination are not patently absurd in relation to the disability (e.g., mental retardation, autism).

Perhaps more important than indicators of overall achievement is the idea of using formative assessment to increase the effectiveness of teaching and learning. Think of formative assessment as a chance to improve a procedure before it is too late. Applied to the instructional process, it means that by looking at students' performance (on an ongoing basis, adjusting as necessary), you can take appropriate steps to ensure that students can grasp and apply the knowledge being imparted or that they are learning through their hands-on experience. From the students' perspective, formative assessment gives them feedback they need in order to know if they are on the right track.

Obviously, if formative assessment is helpful to nondisabled students, it will also be valuable, perhaps more valuable, to students with disabilities.

Here are ways to use assessment effectively with students with disabilities.

<table>
<tr>
<td>

Informal classroom assessment, together with constructive feedback, can be very helpful to the student with a disability.

</td>
</tr>
</table>

Stay on top of student progress; don't wait until it's too late to discover there is a problem.

- Collect information about progress frequently. As students with disabilities carry out projects, be aware of how they are doing. Don't wait until the end of a project or preparation of a report to find there is a problem—informally check (and perhaps have the student chart) his or her progress. Focus on areas of continuing need and build on strengths by acknowledging the student's accomplishments, however small they may seem.

Be sensitive to students' reactions to the process of assessment.

- Be sensitive to the student's reactions to assessment in general. Many students with disabilities are apprehensive about assessment. They are afraid to fail and they rarely see errors as opportunities to learn. While these statements apply generally to most students, they are even more likely to characterize the reac-

tions of a student with a disability who is placed in a regular science class. Try to take the sting out of assessment by casting it in the light of a chance to "share what you know." As much as possible, avoid unhelpful criticism and replace it with positive reinforcement and suggestions ("You are starting to get the idea, now let's try . . .").

- Assure fairness in assessment. Fairness is not served by insensitivity. Central to the idea of fairness is being aware of ways that the disability and the form of assessment can interact in a way that places the student with a disability at a disadvantage completely separate from the content being tested. If it takes a student longer to do his or her homework or complete a classroom assignment, it is probably due to the disability and not to laziness or lack of desire; consequently, there is every reason to expect that he or she will need extra time to complete a test.

Similarly, if a student uses assistive tools for learning, such as brailled materials, recordings, or special equipment, it is appropriate that they be available in a testing situation. On an as-needed basis, regular science teachers should feel free to request help in assessment from the special education staff. Tests may be administered orally (instead of written) and administered in a different setting by someone familiar with assessment procedures without compromising the integrity of the assessment.

- Conduct multiple forms of assessment as indicators of progress. Written, oral, and performance assessments measure different dimensions of learning. A student who has a disability that affects him or her along one dimension may be able to demonstrate mastery better using a different form of assessment.

There is increasing use of portfolio assessment in science to augment more traditional testing. Whereas testing involves a sampling of a specific body of knowledge through questions presented to the student, portfolio assessment represents an accumulation of varied evidence about what the student has accomplished

Assure that fairness is present in all assessment in which the student with a disability is involved.

Making special arrangements for the student with a disability does not compromise the integrity of the testing situation—it ensures it.

Conduct multiple forms of assessment, including written, oral, and performance assessments.

Along with performance assessment and more formal testing, portfolio assessment motivates students to do their best.

and knows. The sampling/testing model is appropriate when standardized assessment and norms are involved, but portfolio assessment is more helpful in gauging the overall quality of a student's work. Portfolios of a student's products, collected over time, can inform the teacher (and others) about the range and depth of the student's accomplishments. A portfolio can also act as a powerful motivator because it provides tangible evidence of growth—encouraging the student to excel and go farther than the minimum performance required to complete a task.

A REMINDER

How you interact with the student with a disability will have an important, probably lasting, effect. E. C. Keller, a science educator with considerable experience in teaching science in classroom, science center, and field settings, has had a lasting impact on many students with disabilities. He and his colleagues state:

"Our experiences have taught us two lessons about mainstreaming: (a) the teacher is the 'role model' for behavior toward the disabled student—his attitude and how he reacts, communicates, and assists the disabled student will be the model for his students, and (b) the disabled student must offer advice, constructive criticism, and assistance to the instructor and his peers regarding his needs."[2]

NOTES

1. See note 1, page 9.

2. Two useful manuals are available directly from E. C. Keller, Jr., at West Virginia University, Morgantown, WV 26506-6057. Several of the case examples in Chapter 4 reflect some of the ideas presented in these two volumes.

   Mainstream Teaching of Science: A Resource Book, E.C. Keller, Editor, 1983.

   Testing Physically Handicapped Students in Science: A Resource Book for Teachers, Harry C. Lang, Editor, 1983.

Chapter 4

Reducing Barriers with Alternative Strategies

The educational process is complex and varied across settings, grades, and subject areas. Nevertheless, instruction and learning have some things in common in all these situations. By focusing on these common characteristics, we can begin to plan effective strategies that will enhance the participation of students with disabilities in science classes. In terms of the teaching and learning of science, there are three process dimensions: teacher-centered instruction, student-centered activity, and partners and team cooperation.

Teacher-centered instruction (pages 39 to 85)

Broadly speaking, teacher-centered instruction pertains to all those activities in which the teacher is necessarily the responsible person or the center of instruction and student attention is to be directed toward him or her. It involves the wide variety of things teachers do, with special attention to the effect that the presence of a disability may have on each activity. Included are

- Assuring that the student with a disability has access
- Using support staff and other students in appropriate ways
- Planning and structuring lessons and activities
- Providing science information
- Using multiple modalities to convey information
- Gaining the students' attention
- Giving directions
- Making assignments
- Evaluating student assignments

Student-centered activity (pages 86 to 112)

Student-centered activities are those in which the student with a disability is the key player—either in doing things on his or her own or responding or reporting to others. Included are

- Communicating needs
- Gathering information
- Obtaining and using materials, equipment, and specimens

- Using computers

- Responding to others and reporting findings

Partners and team cooperation (pages 113 to 131)

This dimension of the educational environment is concerned with those learning activities that are jointly undertaken, where responsibilities are shared among students (with and without a disability) and cooperation is essential. Included are

- Organizing partners and teams

- Sharing technology

- Carrying out assignments as a team

ROUTINE ACTIVITIES IN SCIENCE CLASSES

In all schools three routine activities typically occur in science classes, and students with disabilities should be expected to participate in them. They include attending regularly, following a schedule, and accepting responsibility. Let's examine each of these general concepts and the possible implications arising from the presence of a disability.

Attending regularly

Attendance is fundamental for keeping up with the pace of study and the acquisition of knowledge and experience. While all students with disabilities should attend regularly, any student can experience recurring illness that makes attendance problematic. Some disabilities may have associated health conditions that interfere with regular attendance. These include vision impairment from diabetes and physical impairment from disorders that are health related and chronic.

Schools have an obligation to serve students with such disabilities, even if the disease is terminal cancer or AIDS. Of course, their out-of-school education will be very different from that conducted in the classroom, the science center, and field locations. Nevertheless, the science teacher can help to maintain the student's link to the classroom by suggesting paired activities that can be shared with peers whose attendance is not a problem. If the technology is available and the students are capable, the hospitalized or home-bound student and in-school students might exchange information on current topics under study using computer modems. Alternatively, the phone and drop-in visits (in noncontagious cases) are reasonable alternatives.

Following a schedule

Schedules are a big part of school routine. They are not ordinarily big parts of most students' lives outside of school and perhaps are even less so for students with disabilities. Students with some disabilities may find it harder to get around so they may need extra time between classes. Others may find it hard to meet a schedule for turning in a paper or project because working on that assignment (and other concurrent assignments) may take them longer than the average student. Others may simply forget or get mixed up as a symptom of their disability.

The science teacher should determine whether the student's inability to follow a schedule is beyond his or her control or whether it is a matter of training or motivation. One strategy that can help the student with a disability keep up is to use a unit calendar to list when every activity is to be completed during the study of a science unit.

If the student is often late to class, opening remarks may need to be repeated. Teachers who expect that a student will be late in arriving, or will need to leave a bit early, may find it best to rearrange their teaching pattern so directions for homework assignments are given when all students are present. Other students, such as those with asthma, cystic fibrosis, or hemophilia, may need more time for task completion due to chronic illness and lost school time.

Changes in routine will introduce a lack of consistency, so some thought should be given to this when creating day-to-day lesson plans. For example, if a field trip is being planned and it is essential for all students to be punctual for the bus, the science teacher can chat privately with any student that has been erratic in adhering to a schedule to be sure he or she arrives on time this time. It is also important to discuss the issue with the student's family.

Accepting responsibility

Without exception, all students, including those with disabilities, should be expected to

- observe safety rules in the science class or science center

- show proper care of equipment and living creatures under study

- clean up after themselves

Students with severe disabilities may find some aspects of these responsibilities, such as cleaning up, more difficult than others, but with teaming arrangements, these difficulties can ordinarily be surmounted.

The science teacher must pay special attention to the task/disability relationship and plan ahead to avoid undue hazards or expecting impractical things of the student.

FINDING YOUR WAY IN THIS CHAPTER

The chart on the following page lists each of the topics outlined on pages 35 and 36. This chart is your key to using the strategy information in Chapter 4.

A reminder: The first page for each topic shows check marks next to each of the disabilities that *may* present special problems related to that topic. If there is no check mark, by and large, students with that disability should function within the range of performance expected of students without disabilities.

Look at the first topic on page 39, Teacher-Centered Instruction/Assuring Access. You'll see that only two of six disabilities are likely to present a significant problem for this topic. Discussions of alternative strategies for each disability are provided on subsequent pages in the order shown by the highlighted areas. On page 40, note that the commentary for Physical Impairment applies to grades K–12. Also note that, for the first case example, the suggested strategy applies to the classroom (icon of a teacher's desk with chalkboard) and to the laboratory or science center (icon of a microscope). The second example applies largely to out-of-school activities or field trips (icon of a school bus).

As you use this chapter as a source of inspiration and ideas, you may think of other strategies you have used successfully that would be appropriate in a particular situation. We encourage you to make notes about these strategies in the space that is available on most pages.

KEY TO THE ALTERNATIVE STRATEGIES

HI = Hearing Impairment   PI = Physical Impairment   VI = Vision Impairment
LD = Learning Disability   SI = Speech Impairment   ED = Emotional Disorder

Symbol Key

K–12 Describes the applicable span of grades for a strategy.

Indicates an activity that takes place in the classroom.

Indicates an activity that takes place in the science center or lab.

Indicates an activity that takes place outside of school.

ASSURING ACCESS

hearing impairment
physical impairment
vision impairment
learning disability
speech impairment
emotional disorder

Gaining access to the educational program is essential. The Americans with Disabilities Act *requires* reasonable access and accommodation, so an obvious first activity is to examine the instructional environment in light of the functional needs of the student with disabilities.

To do this efficiently, it's a good idea to have the special education teacher participate in a review of the physical environment. Together, make an evaluation before the student attends class. Questions to consider are

Can the student get to the room?

Can the student enter the room?

Can the student move within the room?

Can the student reach work surfaces, equipment, and materials?

Is the seating appropriate?

Is the lighting adequate?

Will anything present adversely affect the health or safety of the individual?

Some suggested strategies follow.

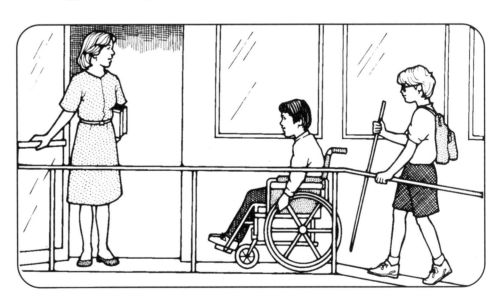

Physical Impairment K–12

In evaluating access for students who are physically disabled, the principal consideration is how well a wheelchair user can maneuver and carry out tasks assigned to other students. Note particularly the width of doorways and aisles in the room (where the wheelchair will be located), clearance under tables, and access to supplies and equipment. Barriers to any of these can impede the wheelchair occupant, but only the doorways would ordinarily require a construction change. Students who use crutches will need a place to conveniently store them so they don't become a hazard for other students. Access also means access to equipment that can be used effectively.

Alternative Strategy

Raise tables onto wooden blocks. Change aisles by relocating desks and chairs. Lower the stored supplies and equipment, or simply give them to the student as needed.

Case Example

Freddy is not able to use his hands in a precise way due to his cerebral palsy condition. Other students can use the regular keyboard on the classroom computer, but Freddy can't hit the keys accurately. For the present, Freddy is provided with head gear with an attached pointer that he can direct to the keys one at a time by nodding his head. In the long run, Freddy will need adaptive equipment that allows him to type more quickly, such as a template that fits over the keyboard or voice-input control.

Access to buildings and transportation outside of school often involves working around barriers. Advance planning is vital.

Alternative Strategy

Be sure an aide, family member, or helping partner is present for any unusual field trip site. If the regular school bus is unable to accommodate a wheelchair, arrange for a specially equipped van-bus to come along.

Case Example

Kristen and her class are on a trip to the airport. By prior arrangement, a special wheelchair (used at the airport) makes it possible for her to enter a plane with her classmates.

Vision Impairment K–12

Access barriers are generally not a problem for vision-impaired or blind students in science classes. However, it is *very* important to a) find out where things are located in the learning environment and b) identify hazards, and, wherever possible, eliminate them.

Alternative Strategy

As a routine, check the instructional environment to be sure it is ready for use by the blind student. If there is a significant change (say, in the placement of seating), be sure to tell the student of the new situation. On field excursions, be sure to assign a buddy or have an aide along to assist.

Case Example

A new shipment of supplies has arrived and they are temporarily placed in an aisle near the door. Edwina, who is blind, knows this room and expects a clear walkway. Because they represent a safety hazard, the supplies are removed before her class meets. In the laboratory, Edwina receives an orientation to the work and supply areas. For several weeks, she is aided by a partner as she learns where things are located.

| hearing impairment |
| physical impairment |
| vision impairment |
| learning disability |
| speech impairment |
| emotional disorder |

USING SUPPORT STAFF AND STUDENTS

From time to time, students with disabilities may have a need that can best be met through direct assistance. However, it is not a good idea to develop dependency on external help but rather to build self-reliance as much as possible.

Intervention by others on behalf of the student with a disability is appropriate for at least one of the following reasons:

- to orient the student or walk through a procedure to be learned

- to escort a student where hazards or barriers exist

- to facilitate communication that otherwise would be lost or incomplete

- to administer tests or otherwise monitor activities when appropriate

- to provide technical consultation or supplementary instruction (by either staff or students)

- to accompany a student on excursions outside the classroom

In addition, partnering students encourages sharing of skills and provides an opportunity for mutual learning. Cooperative learning goes a long way toward the reduction of attitudinal and environmental barriers.

Hearing Impairment K–12

Perhaps the most important forms of human assistance that students with hearing loss may need are a) interpreter services, b) note-taking services when the student relies on lip-reading, and c) partnering when instructional activities involve audible cues in a significant way.

Interpreters may be paid professionals or school staff or volunteers who can sign, and they may be needed either to facilitate inward communication (e.g., "hearing" the discussion groups) or outward communication (where the hearing-impaired person's speech production is nearly unintelligible).

Alternative Strategy　　In the early stages of science class, for a week or two at least, try to arrange for volunteer support in helping the hearing-impaired student adjust to the class routine. In particular, focus on optimizing the student's skills in note taking and in expressing him or herself.

Case Example　　Riga, who is hard of hearing, is coming to regular science class for the first time. To integrate Riga smoothly, one student makes duplicate notes for her. An interpreter comes in each week so that he, the teacher, and Riga can review lesson materials and plan for the week ahead. Also, for at least 10 minutes a day, the special education teacher or aide, who signs, comes in to help Riga stay abreast of the assigned classwork and clear up any misunderstandings from the day's lesson.

Physical Impairment **K–12**

The need for human assistance for persons with physical disabilities typically falls into one of the following categories: a) mobility, largely in locations not designed for wheelchair access; b) manipulation, such as of items that are out of reach or not easily controllable; c) transportation, in situations in which vehicles are not designed for access; and d) note taking, to the extent that upper extremities are severely involved.

Because the term *physical disability* includes such a wide range of impairments, both in type and severity, many students will have needs in only one or two of the areas above or perhaps only in some very limited way not mentioned.

In students with severe disability, stamina and strength are likely to be limited. Some students may require assistance with bodily functions, although this is often handled by someone other than the regular (science) teacher.

Alternative Strategy

Be certain to have the student *tell you* when he or she anticipates a need for assistance. This can be accomplished by a pre-arranged signal between the student and teacher, so other class activities need not be disturbed. If appropriate, provide the assistance, but also provide positive reinforcement when the student shows he or she is able to do something unaided.

Case Example

Ingrid has difficulty holding the science text. Working with the vocational education staff, the teacher has a device made that allows Ingrid to read a text that rests on the arm of her wheelchair. A community volunteer is contacted to act as Ingrid's aide, to accompany her for a community-based project she has elected to do in the area of ecology (investigating pesticides in local gardens).

Vision Impairment **K–12**

The vision-impaired student will most likely need assistance in the following areas: a) conversion of instructional materials, notes, or tests into an accessible form; b) translation of pictorial, graphic, or real but untouchable displays into a meaningful form; c) identification and differentiation of items where touch will not discriminate; and d) orientation and mobility in unfamiliar situations.

Alternative Strategy

Differentiate one-time needs (such as orientation to a new setting or identification and labeling of objects) from recurring needs (such as providing visual descriptions of objects out of reach or transcription of chalkboard notes). Build self-reliance regarding the former needs and provide efficient, reliable assistance for the latter. For example, self-reliance will follow if the student is oriented to where things will be kept and gets to handle the objects that will be used. Ongoing assistance may be needed for converting materials from a visual to a tactile format; volunteers can be used for this purpose.

Case Example

Jessica has learned Braille. To help her with schoolwork, the special education teacher collaborates with the regular teacher to have Jessica's science lessons brailled ahead of time. As part of a science project, the special education teacher helps Jessica make her own Braille labels and helps her attach them so that she can identify her own project materials.

Learning Disability

The student with a learning disability may need various types of help, depending on the specific nature of the disability. Students with a processing disorder affecting auditory language may need things explained more thoroughly or require recorded information to allow repeated listening. Other students with visual processing disorders may benefit from lessons that are multimodal rather than simply print-based. Both types of students would find it advantageous to have science content presented in small, incremental steps.

Alternative Strategy

In addition to various forms of help from special education personnel (e.g., for administering extended-time tests, for developing lessons using different modalities, for helping students learn study skills), regular help from a peer or cross-age tutors is often useful (e.g., for explanations of directions or learning of discrete facts).

Case Example

Eduardo has a visual processing disorder and, to a lesser extent, an auditory processing disorder. His graphic skills are good and he likes to use his hands. As part of a cooperative-learning team, which is being guided by a cross-age tutor, Eduardo helps to design and construct the team's science exhibit. When reading tasks are assigned, the teacher either allows more time or has some of the more complex lessons recorded on tape by a volunteer.

Speech Impairment K–12

Intervention takes several forms for students with speech impairments. Where speech is possible but not intelligible or articulate, speech therapists play an important role. However, they are not normally available in science-oriented instructional settings. In this situation, help may be forthcoming from peer or cross-age tutors and from special education staff.

Alternative Strategy

Instead of oral reports, allow the student to prepare written reports and use peers to read them aloud. Alternatively, a computer could be used or a poster or pictorial presentation could be made.

Case Example

Filipa's speech is understandable to a limited circle of friends but isn't clear enough to be used for presentations. One of her friends is a cross-age tutor in the science class. This tutor serves as an expressive interpreter when Filipa presents her science project to the class.

On a field trip, the speech-impaired student may want to ask questions but is reluctant to do so because it would bring attention to himself or herself.

Alternative Strategy

Rather than have a student with a speech impairment leave his or her questions unanswered, have a buddy ask questions on the student's behalf.

Case Example

Billy Ray cannot speak. He uses a spell-board to indicate his needs. Fred has "interpreted" Billy Ray's expressed needs before, so he helps Billy Ray ask questions on a field trip.

Emotional Disorder **6–12**

It is not uncommon for the student with an emotional disorder to go off-task or overreact to a stimulus. When this takes place, productive learning is not likely. To help prevent these off-task behaviors, call on other people to help. Of course, the focus of the intervention should be to diminish the frequency of the off-task behavior. Take care that the student doesn't see the intervention as a reward for negative behavior or something to be committed routinely.

Alternative Strategy

Set up a contingency reinforcement schedule involving an adult volunteer or a cross-age tutor. Use the special education teacher to work out the contingency plan and to help introduce the intervention.

Case Example

Paula is under medication for mood swings that play out as magnifications of stressful events in the classroom. The medication interferes with Paula's concentration, so she drifts at times during class. Calling on Paula for a response at this time is not advisable because she feels cornered and gets defensive. Use of a positive example in which Paula and a friend or classmate are mentioned will likely gain her attention. To keep her productively involved, the science teacher involves an adult volunteer for field trips and for classroom activities and a peer buddy for science center activities.

Teacher-Centered Instruction

STRUCTURING LESSONS AND ACTIVITIES

hearing impairment
physical impairment
vision impairment
learning disability
speech impairment
emotional disorder

Good teachers are good communicators. They are able to adapt their lessons and activities to accommodate the readiness and the limitations of their students. For many students with disabilities, little adjustment is necessary, while for others an adjustment may be needed in particular situations. Some students with disabilities will take longer to read and execute assignments for reasons that have nothing to do with intellectual or motivational factors. Depending on the disability, adjustments in lessons may involve one or more of the following:

- preparing pre-lesson materials to prepare the students with disabilities

- modifying the form of the materials so they can be used through a different modality

- using more than one modality to present materials and especially more hands-on involvement

- shortening lessons or having more frequent intervals of alternative activity

- involving partners or teams in carrying out tasks

- simplifying student responses required or increasing the time allowed for response

- changing teaching techniques in appropriate ways

For students with disabilities, your question is "How can I present this material to these students so they can learn from it and succeed?" Varied trials, experience, and frequent communication with these students will be the best way to learn what works. To get a running start, plan in advance with the special education teacher.

Hearing Impairment 3–12

Since *sound* is the area of potential problem, lesson plans must take sound into account. When you speak in the class, can the student get the message? If there is any doubt, consider supplementing your speech with written material that covers the same information. If the lesson involves audio materials (tape, film, TV, or sounds that originate in science processes), how will the hearing-impaired student be included? Will it be through an interpreter or captioning? If classmates are discussing something, as in a cooperative group, how can the student learn what's happening in time to contribute effectively? Will the same procedure work for a field-based activity?

Alternative Strategy

For each unit of instruction, decide whether a hearing-impaired student is likely to get the message. As appropriate, make pre-lesson notes to share with the student, use captioned media, have an interpreter come in for the lesson, appoint a team member to take notes or communicate with the student, or arrange for the use of adapted or special purpose equipment.

Case Example

Because Jeanine is deaf, the science teacher is careful to provide notes on the main points that will be covered in his lecture. During question-and-answer sessions in class, the teacher repeats what students say so that Jeanine can lip-read. By arrangement with the school district's media department, he uses a captioned film for a lesson on photosynthesis. On a field trip, he alerts the host staff at the arboretum, and they arrange for an interpreter to help Jeanine.

Physical Impairment K–12

In structuring science lessons for physically impaired students, the principal concern is to ensure active participation. Changes in lessons in the classroom are likely to be minimal.

Alternative Strategy

If classroom activities prove difficult, have a peer help out. Preferential seating is important for visibility during demonstrations.

Case Example

Dixon has muscular dystrophy and is weak. He sometimes slumps over in his wheelchair, and he has little motor control over manipulative objects. Dixon is seated near the door and the teacher's desk. When demonstrations occur, he has a front row seat. When lab activities are scheduled to take place, he is visited by a volunteer buddy (from a higher grade) who gives him a helping hand.

Science center exercises and field excursions should be reviewed and adjusted so that physically impaired students can take part.

Alternative Strategy

When you plan your lesson, decide if an alternate activity in the science center or site for the field trip could make it possible for the student to be included. Have a buddy or adult volunteer assigned if necessary.

Case Example

To include John, a paraplegic, a planned shoreline ecology trip is shifted to a large aquarium. By arrangement, the aquarium host shows slides from a seashore. In the science center, a plan to demonstrate forces and weights using students of different weights is adjusted to take John's wheelchair's weight into account, permitting him to participate.

Vision Impairment   K–12

As you plan lessons, be sensitive to the need for explaining or transforming (into another modality) elements of the lesson that are vision dependent. Many things, such as writing on the chalkboard, can be communicated simply by saying them aloud. Other things, such as illustrations, or objects that are very small, far, or indistinct, must be described in such a way that must-know information is conveyed to the student. Technology can help simplify learning for many vision-impaired students.

Alternative Strategy

For students with low vision, explore the availability of monoculars (such as small telescopes) that let them see the chalkboard, or CCTV (closed-circuit television) image magnifiers that make text and small objects visible. Be sure the student is seated favorably in the room. Make large-print materials available whenever possible. Have models available, if possible, to supplement graphics in a tactile format.

Case Example

Roger has only partial vision, but it is enough to allow him to see and manipulate most classroom and science center materials by using the school's portable CCTV image magnifier. To get the clearest images, Roger inverts the screen, showing white on black, which produces the contrast he needs. Relatively few lesson plan changes are needed for Roger.

For blind students, a combination of Braille and technology can address many, but not all, of their needs. Readers and partners can be extremely helpful in filling gaps. Lesson planning should be done well in advance so suitable arrangements for help can be made.

Alternative Strategy

Have lesson materials brailled ahead of time. Do oral translations of visual media used in class or of objects seen in science centers and on field excursions. Use "talking" (voice input) equipment whenever possible.

Case Example

Calvin is able to use a talking calculator and Braille. With these, he handles his classwork well and in the science center does calculations using data his lab partner supplies. Calvin charts his progress using strings and thumbtacks on cardboard.

Learning Disability　　　　　6–12

The special education teacher will tell you what specific needs of the learning-disabled student should be taken into account as you structure lessons and activities. You may find that little adjustment is required beyond extra time, clearly stated or displayed instructional material, and a balance of hands-on and written study. In addition, you can probably help many students by incorporating cross-age or peer tutoring for complex subject matter. You can also look for ways to chop the science unit into smaller bites or to emphasize hands-on learning. In addition, it's a good idea to check to see whether the student can verbally explain what is to be done to determine his or her level of comprehension.

Alternative Strategy　Make a complex homework assignment due in two or three days rather than one. Have the learning-disabled student describe what is to be done in the assignment and report to you on the first day to discuss his or her progress. If he or she is stuck, arrange for help as needed.

Case Example　Alice's work reflects a disorganized approach to study habits and class work. In a unit on "plant life near our school," the teacher pairs Alice with an observant student who is to collect specimens. Alice will classify and display them. Before they start, the teacher gives Alice a brief side-lesson on classification using illustrated materials contained in a library book. Alice is to make all the initial sorts and classifications; her partner is to help by gathering samples and checking her work.

| hearing impairment |
| physical impairment |
| vision impairment |
| learning disability |
| speech impairment |
| emotional disorder |

PROVIDING SCIENCE INFORMATION

Students are exposed to science information in various forms. Much of what they learn is reinforced by printed materials or other media chosen by the teacher. A substantial amount is transmitted orally by the teacher or other knowledgeable person, and a share is acquired through direct experience. Direct experience, guided by the knowledgeable adult, is often the most enduring and meaningful form of learning. Students with disabilities can benefit from each of these ways of gaining science information.

Any classroom, science center, or field activity that puts the student with a disability in touch with a potential role model is especially helpful. For example, a trip to a local business where science is being applied will be particularly worthwhile if the organization happens to employ a person with a disability in a scientific capacity and that person can speak to the class and show what he or she does on the job.

Flexibility and a willingness to use alternative means of presenting science information provides balance for all the students but is especially helpful to the student with a disability. The challenge to the teacher is in knowing when an information channel can't be used effectively (due to the disability) and coming up with an alternative that still provides the essential information. Creativity, coupled with awareness of technological and human resources, can make a tremendous difference.

Hearing Impairment   **K–12**

The use of objects, pictures, gestures, or pantomimes will enhance oral delivery. Position the hearing-impaired student so that he or she can see both your mouth and any objects being displayed. Supplement oral explanations with written material, and supplement written material with demonstrations or illustrations. If the information is critical and must be received completely to be understood, use an interpreter and/or a note taker. Check frequently to be sure the student has a clear understanding of the information being presented.

Alternative Strategy   Use technology and texts to provide the deaf or hearing-impaired student with additional experiences that supplement or clarify classroom presentations.

Case Example   In a unit on ecosystems, the teacher gives an oral overview and uses the chalkboard to list and define new terms. After introducing main ideas, she has all students read a handout she has prepared. She guides Brian, who is deaf, in using the handout to study a videodisk about selected ecosystems. She follows this with an assignment that pairs students in identifying and describing ecosystems in or near the school grounds.

Physical Impairment 6–12

Unless there is a secondary disability, the physically impaired student usually does not experience difficulty in getting information in the classroom. You may have to make some adaptations when hands-on activities or mobility are important to the learning process. This is most often the case in the field or in a science center. Adaptations may include providing direct assistance, such as a note taker for a student with cerebral palsy, or allowing for a different form of experience gathering, such as using a computer.

Alternative Strategy

When information gathering involves a physical action that the impaired student cannot perform, use a different type of experience that gets at the same information. For example, use data collected by others instead of requiring that it be gathered directly.

Case Example

In a unit on acid rain, Erica, a student with physical disabilities, cannot collect water samples. While her partners do so, Erica is busy with the computer, which she enjoys using. She logs on to a public-access network and contacts a student in a state 1000 miles away. She arranges for that student to get local water samples and send them to her collect, safely wrapped in a clear plastic container. Because of her contribution, Erica's group is the only one that obtains comparative data for the class project.

Vision Impairment **K–12**

Information gathered through the eyes is ubiquitous and naturally a part of most things sighted persons do. Vision-impaired or blind students cannot rely on their eyes to gain information. As you talk to your class, ask yourself, "How is this relevant to the vision-impaired student? Can this student use this information to improve his or her science understanding, or is it being learned as a curriculum requirement?"

For example, concepts of light refraction and the effects of air-to-liquid (or glass) transitions of light rays are not directly useful to a blind student. However, because this principle has important applications, it is worth finding another means of demonstrating the effects of refraction on wave energy.

Alternative Strategy Whenever possible, transform visual information into tangible information and, if appropriate, print information in enlarged print or Braille or use audiotape. If appropriate, involve a student partner in the collection of information.

Case Example In demonstrating refraction to the science class, the teacher uses a special circular refraction tank for an air-to-liquid demonstration.* She also makes a ray box with three parallel slits to let in sunlight, together with a magnifying glass, to show the refraction process with glass. For Greg, a blind student, she has his partner affix toothpicks along the line of the entering and emerging rays. With a protractor (edge notched at 10-degree intervals), Greg measures the angles each time the variables change.

*See NADA Scientific in the Resources chapter (p. 146).

Learning Disability K–12

Without intending to do so, many students with learning disabilities scramble incoming and outgoing information—printed, oral, or both. While there are many explanations for this, the science teacher must be aware that it does occur and take steps to minimize its impact.

The three most important steps are to 1) present information in single, incremental chunks; 2) supplement the main information channel (e.g., speech, print) with reinforcing information (e.g., graphics, gestures, demonstration); and 3) ask clarifying questions and (informally) have the student describe his or her understanding in simple terms, perhaps using familiar examples.

Alternative Strategy

Pace instruction carefully to ensure clarity. Use segmenting techniques, such as lists on an overhead projector, illustrations, and frequent question-and-answer sessions. Monitor progress and adjust presentations accordingly.

Case Example

Peter has difficulty processing oral information. Given four steps of similar nature, he is almost certain to mix them up. As each step is stated by the teacher, she gets Peter to write it down. She then illustrates the steps using hand and arm gestures to mimic physical changes as they occur. Next, she confirms, through questions, that Peter understands.

Emotional Disorder K–12

As you provide information to the student with an emotional disability, be aware of how he or she is functioning that day. For example, if the taking of medication has affected the student's concentration, consider whether there are adjustments you might make to accommodate him or her. Similarly, if the student is showing symptoms of stress, it would not be a good time to place him or her in the spotlight with a lead role or activity.

Be especially aware of the effects that a change in routine can have on the student with a disability. Offset any surprise about the change in routine by informing the student well in advance so that his or her attention can be given to constructive planning for the new activity.

Alternative Strategy

Use time-out sessions to cool off disruptive behavior and as a break if the student needs one for any disability-related reason. Once the time out has been completed, be sure to bring the student back into the group in a non-attention-getting manner. Make sure the student does not lose out on information provided during the time-out session.

Case Example

Mark has sudden and unpredictable mood swings. Knowing this, early in the school year the teacher works out a signal procedure with Mark that tells him to take a break. The signal is a simple hand signal: by holding up two fingers the teacher has signaled the student to take two. At these times Mark is to change to a different project (not do nothing), an unpaced, no-stress activity, such as reading. During science center and field activities, Mark has a buddy assigned whom the teacher tries to keep nearby. At all times, the teacher treats the situation with calmness and places the emphasis on making sure that Mark does not miss out on the information being conveyed.

USING MULTIPLE MODALITIES

| hearing impairment |
| physical impairment |
| vision impairment |
| learning disability |

speech impairment

emotional disorder

Students with disabilities often flourish when teachers use multiple modalities or channels that involve the different senses, supplement oral and written presentation with media, and blend active and passive learning activities. In science, students must learn to be observant, to control variables, to manipulate equipment, to collect and assimilate information, to interrelate facts and events in causal ways, and to report findings and conclusions. Students with disabilities can do most of these things well if adaptations are made. For example, in lieu of participating in hands-on experiments, a quadriplegic student might be able to control a computer simulation. Students with disabilities are especially successful in a multiple-modality environment because it best affords them the chance to acquire through one modality what they couldn't gain through another. It also provides them with exposure to experiences that may have lacked previously.

The science teacher needs to review his or her style of teaching. Does it depend on one channel, such as sound? Can an alternative, supplementary channel be used? The answer may vary by science topic, but in most instances the answer is yes. The regular science teacher and the special education teacher together can identify areas of student need and map out strategies for multiple-modality teaching.

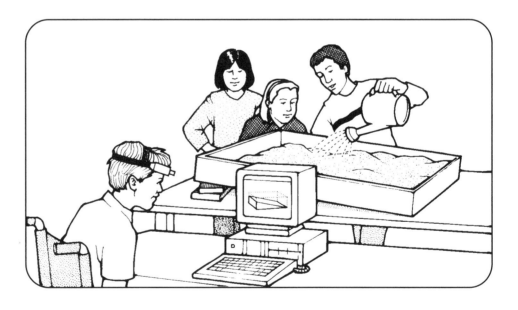

Hearing Impairment

The sensory substitution requirement for hearing-impaired students is apparent. If appropriate, explore the availability of volunteer interpreters if most of the instruction is oral. Alternatively, consider equipment like TDD (telecommunication device for the deaf) devices. Sound-amplification technology and favorable seating may enable the student to hear the teacher, but student comments and questions in group settings may be lost.

Apart from compensation for the disability, the deaf or hearing-impaired student will benefit, just as nondisabled students do, from a rich blend of experiences presented through different media and hands-on involvement.

Alternative Strategy

Using a variety of visual cues (e.g., lip movement, captions for sound media, illustrations, gestures, demonstrations), immerse the hearing-impaired student in an information-rich environment.

Case Example

In a science unit on sound, the teacher usually relates vibrational frequency to audible pitches and uses music and or other sounds to demonstrate. With Jeannie, a deaf student, other modalities such as sight and touch are needed.

For *touch,* a) the teacher uses an open piano, and Jeannie touches the strings as low, medium, and high notes are played; and b) Jeannie places her hand in front of a hi-fi speaker as different instruments and frequencies are produced.

For *sight,* a) the teacher shows overhead transparencies of wave forms; b) the teacher shows the relative rapidity of vibration with taut and semi-loose large rubber bands; and c) on a trip to a science museum, the teacher shows Jeannie the apparatus that allows sound waves to be visually displayed.

Physical Impairment K–12

Physical disability can interfere with instructional modalities that involve the sense of touch and fine or gross motor performance, including strength, agility, and endurance. Of course, access to the instructional setting and to equipment or objects to be manipulated must be possible. Hands-on activities, important as they are, may need to be examined to determine how they can best be accomplished in relation to the particular physical impairment.

Alternative Strategy Think in terms of alternatives that demonstrate the same scientific principles but that make fewer physical demands. Alternatives may include plasticware, beakers with handles, lap boards for writing, and other special tools.

Case Example Willy has muscular dystrophy and is weaker than his classmates. In a class on force, levers are used as a demonstration and later in experiments. Willy is able to participate when the power advantage becomes favorable. Later, in a lesson using pulleys, he is again able to lift a weight he could not normally handle.

Vision Impairment

The level of vision loss will be central to deciding how the science teacher can present multiple-modality instruction. Particularly important in the classroom are the graphic images that are in texts, transparencies, films, and TV. In the science center, chemical reactions and microscope use can prove difficult. During field activities, the objects to be observed are often beyond the range of touch, screened from touch by a container, or are in fact intangible.

Hearing can substitute for impaired vision but may have limitations because it requires the use of analogy (e.g., a kettle whistle is sounded when steam is generated). It is therefore dependent on adapted or special purpose equipment.

Alternative Strategy

Have the student use a recorder to play audio-taped texts or to tape class lectures for later listening. Use real objects and models for three-dimensional representations whenever possible. Use raised-line touch (Thermoform) representations.* Tag shapes and relationships (such as distance comparisons) with buttons or other markers. Use technology that is designed for or adaptable to low-vision use. Use Braille or enlarged print as appropriate.

Case Example

Philip has partial vision; John is blind. Both are in a science class studying cells and cell division and reproduction. Both are able to benefit from plaster-of-Paris models of cell structure and cell division. For microscope work and direct observation, the school district acquires a Microvix microscope video processor, which allows Philip to view the slides along with his classmates.** A verbal description is given to John by a sighted classmate.

---

*See American Thermoform Corporation in the Resources chapter (p. 138).
**See NADA Scientific in the Resources chapter (p. 146).

Learning Disability

Risks and rewards are associated with using multiple-modality instruction with learning-disabled students. On the reward side, it affords the student who is weak in one learning modality an opportunity to get information through a preferred modality. For example, a poor reader may be a good listener or good with hands-on activities. However, a number of students with learning disabilities are highly distractible, and such students may not focus on the material to be learned if they are simultaneously bombarded with competing stimuli. In view of this, it's reasonable to sequence alternative strategies so that the student is only being asked to focus on one thing at a time.

Alternative Strategy

Try to think of ways to connect abstract ideas to the experience of the learning-disabled student. Support one modality by following it with instruction that uses another modality. Consider using rhyming and songs that incorporate new vocabulary as a way of making learning fun and effective.

Case Example

In a lesson on heat absorption, there is a discussion about how the rays of the sun cause heat to build up. Bill, who has been diagnosed as dyslexic, clearly did not learn the principle from reading the text. He listens but does not actively contribute to the discussion. The class is challenged to devise ways of testing heat absorption hypotheses. To engage Bill, the teacher has him record the temperature when a thermometer is wrapped with aluminum foil and placed in the sun for five minutes. Next the foil is painted black and another measurement is made in five minutes. Finally, the teacher asks Bill to tell the class which is hotter in the summer, a black macadam street or a concrete sidewalk.

GAINING THE STUDENT'S ATTENTION

hearing impairment	
physical impairment	
vision impairment	
learning disability	
speech impairment	
emotional disorder	

It's easy to take things for granted—like announcing to the class that you want them to stop what they are doing and do something different and expecting them to respond accordingly. It's an action that teachers repeat time after time in the classroom, the science center, and during field excursions. But if the student is hearing impaired, how will he or she know that you want to change the activity? Will the student with a visual impairment be aware that you are making a change? Will the student with a learning disability be able to shift attention in the way you intend? In other words, will the student with a disability catch on or be left out because you failed to get his or her attention?

Getting the attention of the student with a disability is a relatively simple matter, but you do have to set up a practical procedure. Early in the semester, talk with the student about how you think you might accomplish this. Perhaps the student will have an idea or the special education teacher can suggest an effective technique.

Hearing Impairment

Ordinarily, teachers call for the attention of students in the classroom when they want to make an unplanned announcement, say, to redirect attention from a student-centered activity to a teacher-directed activity. Students who are deaf will be unaware of the need to pay attention to the teacher unless a nonvocal cue is used.

Alternative Strategy

Seat the student close to your desk, facing you. Use a hand signal or touch lightly to indicate that you want attention.

Case Example

Meela is hearing impaired. She is seated close enough so that when the teacher wants her attention, she puts her hand on Meela's desk.

Science centers often have workstations scattered around the room. A number of the students will be seated so they do not face the teacher. Because of the seating, step-by-step directions may be hard to follow.

Alternative Strategy

Use an overhead projector to show step-by-step instructions. Mask all the instructions except the one that you want followed next.

Case Example

Leanne and Bill are in the same class. Both are hearing impaired. The teacher seats them where they can see the screen and their work. Using an overhead projector and a transparency showing the lab steps, he pulls the mask (a sheet of paper) down one step at a time. Both students can see which step is next.

During field trips, students tend to spread out. This is especially true if there is a free-time period when students can move about independently, such as within a specific section of an aquarium, arboretum, museum, or park. The teacher needs to gain their attention to have them reassemble. A hearing-impaired student may not be in a position to see or hear the teacher.

Alternative Strategy

Use a buddy system. Have a hearing student accompany the hearing-impaired student. Try to choose a reliable hearing student and have a contingency plan.

Case Example

Pedro is in the fourth grade and is deaf. His regular buddy is Todd, but he is out sick on the day of the field trip. Matthew substitutes. The teacher also tells Pedro a time to return.

Vision Impairment

Posted announcements, whether they are on the chalkboard or bulletin board, are likely to be missed by blind or vision-impaired students. A regular procedure should be developed so each posted announcement is also communicated to the students with vision impairments.

Alternative Strategy

At the time that the announcement is posted, make a copy for the student or make it a point to have it read out loud.

Case Example

Each week, the class assignments are posted beside the door. The day they are posted, Kim receives his own copy, already in Braille.

Occasionally during the course of a science activity, the teacher needs to draw students' attention to an important aspect of the activity that is vision dependent. For example, a teacher might tell students to observe the behavior of fish in a tank as they feed or protect their territory. The student with vision impairment hears the direction but has no way of carrying it out.

Alternative Strategy

For vision-impaired students, set up a closed-circuit television system to magnify small items that need to be observed. For blind students, have another student describe the action as it is observed.

Case Example

Edward sees things best if they are enlarged and have high contrast. To help him see his own experiments, a jeweler's light with magnifier is placed at his work-table.

During field trips, students tend to spread out. This is especially true if there is a free-time period when the students can move about independently, such as within a specific section of an aquarium, arboretum, museum, or park. A vision-impaired student may not be in a position to hear the teacher and may not be aware of the need to reassemble.

Alternative Strategy

Use a buddy system. Pair the student with a seeing student. Have the student with sight describe the things that take place, unless they are accompanied by a teacher's explanation.

Case Example

Charlene can see only at close range. She is paired with Jill, who can see. Jill describes the different kinds of fish in the aquarium.

Learning Disability

Typically, a student with a learning disability must devote a considerable amount of energy to concentrating on the task at hand. When a teacher shifts the focus of an activity in the middle of a lesson, a student with a learning disability may get confused (though not necessarily). If confusion does arise, it may be because the student is concentrating on what he or she is doing and doesn't attend to the specifics of the instructions. Write instructions on handouts, and have students underline key words or directions on activity sheets. Be sure the message was received fully and was understood.

Alternative Strategy   Get the student's attention, then give the instruction on the change of activity. Check with the student to be sure he or she understood. Repeat the instruction if necessary.

Case Example   Ishmael has a problem remembering the sequence of instructions. The teacher announces directions one at a time. She confirms that each step is complete before proceeding.

Science centers have various workstations around the room. A number of the students will be seated so they do not face the teacher. Because of the seating, verbal step-by-step directions may be difficult for students to follow.

Alternative Strategy   Use an overhead projector to show step-by-step instructions. Mask all the instructions except the one you want followed next. Pull the mask down one step at a time.

Case Example   To be sure Philip is on the right task, the teacher monitors his work, revealing the next step on the overhead projector when she sees he is nearly finished with the one shown.

On a field trip, the primary concern for a student with a learning disability is that directions are followed correctly. This includes directions on where to go, what to look for, and when and where to meet.

Alternative Strategy   Provide written directions to the student the day before the field trip. Ask him or her to study them and ask you about anything that is not clear before he or she goes on the trip.

Case Example   A day in advance, Edwin is given a pocket-size list of things to do at the zoo. He is to check off each one he does. The last one is the time and place of meeting for the return trip.

Teacher-Centered Instruction

GIVING DIRECTIONS

| hearing impairment |
| physical impairment |
| vision impairment |
| learning disability |
| speech impairment |
| emotional disorder |

In science settings, students should follow instructions closely. Frequently, the activity is complex. Altering the sequence of steps, the quantities or types of ingredients, or other variance in procedure could cause the activity to fail. It is even possible that harm could come to the student with a disability, to others, or to the specimen being studied.

For students with disabilities, there is a risk that the instructions you give may not be received clearly and completely. For that reason, it is essential that you take extra care in

- *what* you say

- *how* you communicate it

- *whether* you verify that it was received correctly

- *how* you accommodate gaps through alternative strategies

Failed communication often occurs because the sender of the message has misjudged the receiver's capacity to receive it or because the latter has little interest in receiving it. You'll want to concentrate on clear instructions, presented in a format consistent with the student's disability. You'll also need to take steps to confirm that instructions were received as intended. Be prepared to modify directions on the spot as needed.

Hearing Impairment K–12

Hearing-impaired students in a regular science classroom in which the teacher does not sign will probably miss out on some portion of the speech that takes place. The teacher's job is to identify situations that will lead to incomplete information transfer and to use more than one mode of communication.

Alternative Strategy   Use an overhead projector in place of the chalkboard. This allows you to face the students instead of speaking with your back to them.

Case Example   Peter lip-reads during class. After class, he copies the overhead projector diagrams or notes so he misses very little.

During instruction in the science center, the student's attention is directed toward the activity itself. The hearing-impaired student must also have eye contact with the teacher.

Alternative Strategy   As you demonstrate a procedure, deliberately alternate between speaking and manipulating the materials. This allows the hearing-impaired student to look at one thing at a time.

Case Example   Jennifer is learning to use the computer. The teacher first states what needs to be done, then does it. Jennifer repeats each step.

If the field activity takes you to a location where eye contact will be difficult to maintain, consider alternatives so the student can get the necessary information.

Alternative Strategy   Depending on the situation, consider using an interpreter or making notes available to the hearing-impaired student.

Case Example   On a trip to a planetarium, Rob will not be able to see the lecturer in the darkened room. By using an interpreter and a small flashlight to illuminate the interpreter, he gets the same instruction as his peers.

Physical Impairment

In giving directions, the natural tendency is to specify steps that can be performed by physically able persons. You'll need to review these steps to see whether they present difficulties for the physically impaired student. If so, decide whether the steps can be accomplished in a different, modified way.

Alternative Strategy

For steps that the student with a disability may accomplish, meet with the student and explore modified procedures together. Do this in advance of the class in which the steps are to be applied.

For steps that he or she *cannot* accomplish, have the student work with a student partner or aide.

Case Example

Erica has limited finger dexterity. A computer program requires several keys to be pressed at the same time. With the use of, for example, a Macintosh and its Easy Access feature (part of the regular system software), Erica needs to press only one key at a time.

Francine has involuntary muscle movements and cannot use the mouse or keys. She uses a voice-activated peripheral with her computer to duplicate mouse functions.* If the precision of the work is too exacting, she works with a student partner and directs his activity.

---

*See *Voice Navigator* by Articulate Systems in the Resources chapter (p. 138).

Vision Impairment

Handouts and oral announcements are the customary means for giving directions. Students with vision impairments may have little difficulty in receiving oral announcements, and handouts can usually be prepared in advance either in Braille or large print. If the student has access to electronic equipment that can enlarge the image or transform it, the original handout can be used. As directions are given, be sure all aspects of directions that involve sight are explained in detail.

Alternative Strategy

When one step depends on visual cues from a prior step, the student needs some way of knowing when it has taken place (and what it is). If the cue is not tangible, some means must be found to inform the student with vision impairment.

Case Example

Talia is blind. She takes notes in Braille, so she is able to get directions given by the teacher. However, in carrying them out, she is unable to see the results (the directions involve creating visual magnetic fields with iron filings). Several days prior to the class, Talia and a teacher's aide carry out the experiment in the main school office. After covering the photocopy machine with plastic wrap to keep it clean, the filings are carefully spread over the plastic wrap. Two magnets are carefully positioned on the filings. A copy is made each time a different step is carried out. Later these copies are used to make raised Thermoform facsimiles. As she touches each one, Talia is asked to explain the magnet placement that produced the effect.

Learning Disability

Often directions involve complex steps or steps associated with directional attributes (left, right, behind, beside, etc.). The student with a learning disability can get directions twisted as he or she receives them and then performs each step. Whether the directions are written or oral, they should be kept simple and clear, and preferably presented one at a time.

Alternative Strategy

Instead of presenting the directions for a complex activity all at once, present them as a numbered checklist. As each one is completed, have the student indicate that it has been accomplished.

Case Example

Emilio has a reading disability. As a result, printed materials can get mixed up or important information overlooked. To avoid this, the teacher prepares a numbered checklist of steps. As each step is completed, Emilio marks off the number and starts the next. Particularly complex steps are circled. That tells Emilio to confirm with the teacher that the direction is understood before proceeding.

Emotional Disorder 4–12

Students with emotional disabilities often have low frustration tolerance. Therefore, when instructions and directions are given, attention should be paid to a) whether the wording is unambiguous and b) how likely the step is to create a problem during implementation. In the former case, the teacher needs to clarify; in the latter, the student should be told what to do in case a problem does occur.

Alternative Strategy

In an experiment in which the directions are lengthy, try pairing two students. Have them take turns carrying out the directions, each acting as a co-pilot for the other, helping as necessary.

Case Example

Arthur has a low tolerance for ambiguity. Before the experiment, the teacher explains each step. During the experiment, the teacher praises Arthur for completing complex steps. Throughout, Arthur and a second student are working as a team to complete the experiment.

TEACHER-CENTERED INSTRUCTION

Section A

Teacher-Centered Instruction

MAKING ASSIGNMENTS

hearing impairment

physical impairment

vision impairment

learning disability

speech impairment

emotional disorder

Hopefully, the special education teacher has already established a strong link between the home and the school for each student with a disability. In any event, as the student with a disability enters a regular class and participates with nondisabled students, it is important that the family members, as well as the student, are informed early on about expectations concerning homework and science projects done out of school.

Assignments should be tailored, in type, amount, and duration, to the capacities of the student. For many students with disabilities, the actual time required is substantially higher than it is for the regular students, merely because of disability-related barriers and not because of any intellectual failing. To the extent possible, assignments should be coordinated with the special education teacher and the family, as each can be a valuable resource to the student (stopping short of doing the work for the student, of course).

Assignments for long-term projects often involve work with the school or community as well as the home. In such cases, the science teacher needs to provide ample support throughout the term of the assignment and suggest alternative strategies when barriers arise.

Hearing Impairment 6–12

Assignments that have a heavy reading requirement (e.g., many pages to be covered in a short time, or several word problems) will likely be harder for students who were deaf before they learned to speak because of the delay they experienced before they began to pick up language and because of the difference between English grammar and the structure of American Sign Language. More reading time may be needed. Assignments that involve oral data gathering (e.g., polling students to create a chart on heights, with a male-female distribution by grade level) might present a challenge to the hearing-impaired student, though it could be a worthwhile experience nonetheless (e.g., if the polling aspect was eliminated and heights were measured).

Alternative Strategy

To increase accuracy in assignments, have the hearing-impaired student keep a record of any questionable aspect of the assignment. Go over these with the student to provide clarification. If ambiguities arise in the home, have the parents make a note of them.

Case Example

Ingrid has been in an environment where signing is common, but now moves into one where it is absent. Her lip-reading skills are not well developed. On advice of the special education teacher, all assignments out of school are written down for Ingrid. Because her parents can sign, a plan is worked out so that they can help to interpret the assignment if she needs clarification. This procedure works well in an assignment on identifying sources of heat and light energy in the home.

Physical Impairment

Two questions should be considered when giving out-of-school assignments to students with physical impairments: Can they carry out the assignment unaided? Can they generate a report in the desired form describing the outcome of the assignments? In both cases, assignments can be readily screened in terms of the physical demands posed in relation to the disability. If a particular assignment appears to be unworkable, perhaps it can be modified or an arrangement made so that the difficult portion can be directed by the student but executed by a helper.

Alternative Strategy　Plan for assistance with unworkable parts of the assignment. For giving the assignment itself to students with writing difficulty, either have assignments prepared ahead of time or enlist another student to make duplicate notes on the assignment.

Case Example　In undertaking a science fair project, Kelly, who has cerebral palsy, has chosen to use a computer to simulate measurements he could not personally take in the field. For constructing the exhibit at the science fair, Kelly works with a partner. He designs the display, locates the illustrations he needs in the library, and has them photocopied. He also has the help of his sister in assembling the display on the day of the fair.

Vision Impairment

Students with vision impairments may need considerable leeway in how they carry out assignments. As assigned, the nature of samples to be taken or the observations that need to be made might be impractical, though there may be a way to simulate or substitute to achieve the same general objective. Thus an assignment that presumes a classification process involving color, for example, might be accomplished if shape is used in lieu of color and the shape can be distinguished tactually. If the assignment cannot be altered without introducing a safety problem or substantially invalidating the outcome, it will probably be necessary to have a peer share the assignment with the vision-impaired student.

If the vision-impaired student has a way of taking notes in class, taking down assignments will present little difficulty. Otherwise, the teacher, special education teacher, or aide will have to do so and send the assignment home to the attention of the parents or guardians. Advance coordination will be needed so the assignment does not get misplaced.

Implementing the assignment assumes that materials such as Braille or large-print text are available and that the activities are not sight dependent (e.g., making a descriptive log of the birds seen in the next week). Adjust assignments to substitute a tangible task for a visual one.

Alternative Strategy

As assignments are made, try to involve the student with a visual impairment in deciding whether he or she will be able to carry out the assignment independently. Decide whether the assignment can be adapted or whether partnering will be appropriate.

To simplify the task and minimize time and effort in short-term assignments, allow the vision-impaired student to prepare oral reports from his or her own Braille copy. Similarly, present assignments to the student in a form that he or she can use directly.

Case Example

Roger has partial vision. He is able to see print if it is enlarged sufficiently. His assignments are prepared by the science teacher on a copy machine that allows a 156% setting on size. This produces a large-print version from ordinary typing. (Spacing on the original must take into account the paper size on which the assignment will be printed.)

Learning Disability

Two keys to giving assignments to students with learning disabilities are a) to lay out the steps in the clearest, briefest form possible to avoid confusion; and b) to verify that the steps are understood before releasing the student to carry out the assignment.

Alternative Strategy Use check questions to establish the clarity of understanding the student has about the assignment. Use an overhead projector to list tasks.

Case Example Lucy has a reading disorder, so her teacher checks to be sure that an assignment is understood. She asks Lucy, "What will you do first? What next? How will you check your work?" and any other appropriate check questions. If anything is unclear it is revised on the spot. If the assignment involves homework, Lucy is encouraged to check her work with someone at home.

Teacher-Centered Instruction

EVALUATING STUDENT ACCOMPLISHMENTS

hearing impairment

physical impairment

vision impairment

learning disability

speech impairment

emotional disorder

For students with disabilities, evaluation is much more than just a basis for grading. It is a useful form of feedback, informing both the teacher and the student; the teacher improves his or her instruction and the student realizes the progress he or she is making.

Accommodation in assessment is often appropriate for students with disabilities. The most common accommodation is in granting extra time. Another is to allow the assessment to be administered separately, using the preferred (most efficient) modality for the student. This may lead to involvement of the special education teacher either to help prepare the assessment or to administer it.

Apart from testing, which samples knowledge or skill in particular domains, there is value to evaluating the student with a disability in terms of the nature and quality of productivity over time. Portfolio assessments are ways for students to collect examples of their best efforts. A portfolio of accomplishments can be especially helpful for the student with a disability and can help create a sense of pride and competence.

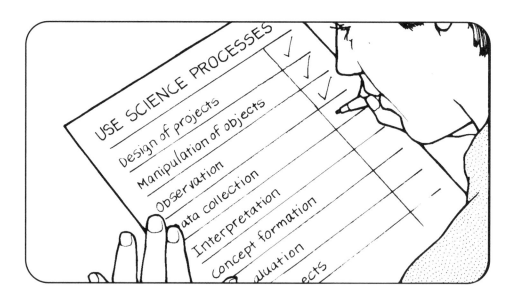

Hearing Impairment

Frequently there is a marked difference in facility with written language between students who have been deaf since birth and those who had sufficient hearing to develop spoken vocabulary as children. Written assessments ordinarily will present more problems for the former than for the latter. An interpreter may be needed for oral assessment, and separate administration may also be appropriate.

As the nature of assessment shifts toward hands-on performance, special accommodation diminishes except when the assessment involves sound in an integral way.

Alternative Strategy

When verbal facility is limited, be prepared to allow extra time for completion of assessments. Design assessments so hearing sound per se is not a requirement to success. If oral production is required and the student's speech is impaired, use an interpreter if possible.

Case Example

In Matt's class, the teacher maintains folders for each student. The folder has two parts: "What I plan to do" and "Here is my best effort." Goals are set by the teacher and Matt, who is deaf, working together. Strategies for assessment are jointly planned between the science teacher and the special education teacher. As Matt progresses, he sees that he has covered many different topics. He takes pride in the work he places in his portfolio.

Physical Impairment 4–12

The nature of a student's physical disabilities will affect how he or she is assessed and evaluated. Oral administration of assessments is appropriate for students with upper-extremity limitations, which implies separate administration of the assessment. Similarly, any performance-based activity inconsistent with the disability will need to be modified or eliminated.

Because students with physical disabilities may have motor problems that mitigate against rapid handwriting or completion of manipulative tasks, additional time may be needed in assessments.

Alternative Strategy   Develop a portfolio of the student's work, both singly and as part of a cooperative group. Orally quiz him or her to establish the extent to which the student contributed to the group-based accomplishments.

Case Example   Bettina has only one arm. She is allowed extra time to complete tasks in the science center. In her assessment, she is credited with having been a leader on a cooperative team because she was in charge of the planning for a science project.

Vision Impairment 6–12

Extreme care must be taken to avoid measures of student knowledge and performance that place vision-impaired students at a disadvantage because of their disability. Thus, in place of assessments such as "Sort these fingerprint samples according to their type: whorl, loop, or arch," modify the task to establish the knowledge using tangible simulations (string glued to cardboard in the three main shapes).

Allow extra time prior to assessments so that brailling or other conversions can be accomplished. Note also that the time needed to read brailled test materials is greater than that needed to read ink print.

Alternative Strategy

Role-play the evaluation exercise or assessment. What needs to be changed to avoid a bias against the student with a vision impairment? Adjust the exercises appropriately, or use technology to compensate for the barrier. Add time for the student to complete the task, and use oral testing freely.

For evaluations that depend on graphic or pictorial content, convert the information to a meaningful format or drop the item as inappropriate.

Case Example

Vihn, who is blind, has been able to use a talking computer for doing calculations in his regular classwork and during assessments.* Instead of using brailled material, he is tested orally for his knowledge of content. In hands-on performance, he is asked to describe and demonstrate those skills that are not solely vision dependent. The assessment takes place one on one. Oral feedback is supplied by a proctor for elements that do involve vision in some way.

---

*See Telesensory Systems in the Resources chapter (p. 140).

Learning Disability

In designing the assessment, take care to defuse test anxiety that can interfere with the validity of the evaluation of a student with a learning disability. Shift the focus from "we will find out what you don't know" to the more reasonable "show or tell me what you have found out so far." Break the pattern of failure that arises from frustration with the pace or the complexity of the assessment. Allow extra time, and try to present assessments that do not generate confusion.

Alternative Strategy

Use portfolio assessment as a way of encouraging the student to do his or her best without anxiety. Have practice exercises to help prepare the student for knowledge and skill measures, and allow extra time.

Consider the use of illustrations by the student as an acceptable form of response to questions in lieu of written responses.

Case Example

Ruthie has a language-processing disorder that makes her anxious about taking tests. She has been part of a study team that has been growing plants in a climate-controlled project in the classroom, and she has kept records in her portfolio throughout.* Because Ruthie expresses herself well, the teacher asks her to explain what she knows about growth factors (light, soil, water, heat).

---

*See American Thermoform Corporation in the Resources chapter (p. 138)

TEACHER-CENTERED INSTRUCTION

Emotional Disorder

Students with emotional disorders are more responsive to positive feedback than negative feedback. They have often heard criticism and may be immune to correction that is presented in a negative way. Tests can be very threatening to students with emotional disorders since they approach them with the expectation of "how many will I get wrong," not "how many will I get right." Portfolio assessment, on the other hand, allow students to be involved in selecting representative pieces of their work and can lead to more enthusiasm and better work.

Pay attention to whether the student is on medication or otherwise may not be functioning at his or her best. In such a case, consider allowing extra time or a deferred test session so that the level of anxiety does not interfere with the assessment process.

Alternative Strategy Have each student accumulate, in his or her portfolio, several examples of work (quizzes, assignments, projects) that demonstrate knowledge of the subject matter or unit of study.

Case Example White Feather tends to withdraw from classroom activities whenever possible. She would rather be "out-of-sight, out-of-mind," and it takes special planning for the teacher to involve her in activities. However, her brightness and knowledge of science are evident in the quality of her work. She warmly accepts the idea of portfolio assessment because she is proud of her work. The portfolio presents a good opportunity for her to come out of her shell. She gets praise for her good work from the teacher in class and from her family at an open house.

Student-Centered Activity

hearing impairment
physical impairment
vision impairment
learning disability
speech impairment
emotional disorder

COMMUNICATING NEEDS

Even though you may have provided adaptive learning tools such as Thermoform versions of graphics, ambiguities may still exist. The adapted instructional materials may not be as precise or clear in their presentation of information. Students should be encouraged to ask questions when they need clarification to grasp the content of the lesson.

As students with disabilities move through school, they need to increase their self-reliance in preparation for adult life. Part of their development involves learning what they can and cannot do, and, especially important, knowing when to ask for help and how to communicate their needs. School, and teachers in particular, typically try to minimize barriers for these students. Too often, they only request that students do things that are easy. While this may be appropriate in some instances, it is not likely to build self-reliance or the judgment to know when and how to get assistance.

Science-related activities present an excellent opportunity for encouraging both of these positive outcomes. It is possible to structure learning activities so that the student with a disability is guided in making choices about what he or she needs as well as communicating those needs to someone in a position to provide help. This can result in a willingness to try something without always expecting success; an increased understanding of what works in the context of the task and the disability; and open, candid communication among the student, his or her classmates, and the science teacher.

Hearing Impairment

Hearing-impaired students can excel at many tasks, but first they must realize that they are supposed to do the task. Too often they don't realize that communication has occurred or they miss out on critical information they need to do a task right. These students commonly encounter gaps in communication, and they should be encouraged to speak out or otherwise signal when an oral statement needs to be repeated, manually signed, or transformed into another medium, such as writing. Acknowledge that you need to know when you have failed to communicate clearly and that you want the student to tell you. For example, if the student sees other students taking notes and the teacher is not directly in view, a question might be raised as to whether some instruction is being missed. Because this may happen more than once, it makes sense to decide jointly on a routine procedure for getting information gaps filled.

Alternative Strategy

Talk with the student about your expectations. Set up ground rules: "I can help you and you can help me." The student is to help by speaking up about his or her needs and making suggestions about what kind of help is needed. Different signals could be agreed upon: "If I touch my shoulder, it means would you please explain it again." An ongoing back-up arrangement could be devised by instituting a buddy system that encourages sharing of notes.

In team settings, develop procedures so the deaf or hearing-impaired student can express his or her communication needs to the other team members, perhaps through a special team helper.

Case Example

Frieda relied on manual signing in her previous school. Now she has entered a mainstream environment and is learning to lip-read. Her science teacher meets with her and together they decide on ways in which their communication can work best. Frieda says that he could trim his mustache so she can see his lips easier. After they talk about ongoing helping procedures, they work out a plan for homework. The teacher stresses that he would like to be sure Frieda understands assignments. Routinely thereafter, the teacher gives a homework assignment and looks to Frieda for confirmation (a head nod) that the message has been received.

Because her speech is not clear, Frieda asks for an interpreter to translate when she gives her first report, which she signs to the interpreter. Shortly thereafter, by prior agreement, she starts to make short presentations without an interpreter. While encouraging her to become more self-reliant, the teacher has also paid close attention to meeting her needs.

Physical Impairment K–12

The teacher will often have a pretty good idea about what physically disabled students should attempt unaided. However, a lack of stamina, or personal needs such as assistance in the bathroom or getting a drink, or unanticipated needs such as a student-initiated activity that requires the assistance of an able-bodied person, may not be so apparent.

Meet with the student and make clear that he or she is expected to do everything possible and is expected to let you know when help is needed. Routinely look for patterns in the requests for help. Repeated requests suggest a need for a long-term solution. No requests may mean that nothing difficult is being attempted and that more communication is needed between teacher and student.

In all cases, strive to develop the student's self-confidence and self-reliance, but let the student know it's not a big deal to ask for help. It is not productive to depend on others when it is not necessary to do so.

Alternative Strategy

If possible, bring to the student's attention a role model with a similar disability. Point out that this individual got ahead by a combination of his or her effort and by asking for help when needed. Together, when new projects or major tasks are undertaken, talk about needs that can be anticipated. Help the student learn what to try independently and when to signal for help.

Case Example

In carrying out a hands-on learning experience involving chemicals in the lab, Kenneth, who is paraplegic, anticipates no problems with anything he can reach. Together, Kenneth and the science teacher examine the tasks ahead. They identify three things that may present problems. One, turning on the gas for a Bunsen burner, Kenneth handles by asking a classmate to do it. Collecting supplies from a storeroom is facilitated by the teacher's placing Kenneth's materials where he can reach them. The third problem, involving a laboratory sink he can't reach because of lack of wheelchair clearance, is temporarily corrected by attaching a hose to the outlet and extenders on the handles. Later the sink is modified by the school's engineers to conform to access guidelines.

Vision Impairment

Blind and vision-impaired students' needs are not always centered on the vision-related problem. For example, blind students may have mannerisms (such as rocking), social problems (such as a tendency to isolate), or may experience difficulty when the activity is not repetitive. It is helpful if they can learn through an efficient routine. For example, they shouldn't have to relearn continually where a familiar object is kept in order to use it. Tell the class to return equipment they use to the place where they found it.

Encourage the student to question statements in which the referent is not clear ("Take it over there"), vision-centered directions ("Write down the order of colors you see as the white light is refracted"), or instructions that will clearly require extra help ("Measure the shadow length each hour through the day").

Alternative Strategy   Develop a red-flag approach. *Red flag* refers to something that can be touched but cannot be understood unless seen. The student can raise one finger or make some other signal when he or she encounters a red flag. Meeting ahead of time can help avoid many red-flag situations.

Case Example   Theo is blind. Using the red-flag approach, he signals in class when a) a student writes his findings on the chalkboard but doesn't say them aloud; b) the teacher demonstrates a task in the science center and says something unclear such as, "Take a measurement at each of these markers"; c) a fellow·student says, "I didn't expect that to happen"; d) the teacher says, "George's specimen looks good; use that as your example."

Learning Disability K–12

Students with learning disabilities may try to draw attention away from their disability in the regular classroom. Because they don't look disabled and often do not draw attention to a difficulty, they sometimes manage to avoid situations in which the disability actually has an impact. Such students may not request needed help. Conversely, some students with learning disabilities may see participation in a group as a good way to have others take on more than their share of the work and may ask for too much help from their peers.

In either case, concentrate on helping the student to differentiate between being stuck and trying to defer or avoid challenging tasks. Initially, make clear to the student what is expected in class and in group participation. Give plenty of positive reinforcement when it is evident that the student is trying things that the disability makes difficult. Also, give praise if he or she asks for help of the right kind; for example, by asking for assistance in completing the assignment and by not asking that the work be done for him or her by someone else.

Alternative Strategy

Together with the student, make up short lists of requests for help that are OK and those that are not OK. Convert the two lists into + and – marks and have the student keep a tally of which he or she does most often. When the "right" list becomes long and the "wrong" list becomes short, replace keeping score with positive reinforcement.

Case Example

Bettina is not good at reading. She avoids it and asks others to summarize or explain things to her in lieu of reading for herself. The science teacher is informed about this pattern by the special education teacher in a conference. With Bettina's full participation, a schedule of reinforcement is set up for the science class.

In an assignment on rocks in which new terminology is introduced (*igneous, metamorphic, sedimentary*), the teacher has Bettina practice the new words (written and spoken) before she is given the reading assignment. When she gets stuck, a reading partner helps at her request. If the two students can't resolve the reading problem, the teacher is asked to help. The teacher does so by using rock samples and helping the students see the relationships and identifying features of the various subtypes.

Speech Impairment

All speech impairments tend to make the communication of needs time consuming. For this reason both students with speech impairments and their teachers may undercommunicate. Try to develop a signal system that shortcuts the requirement for speech. At least part of the time, ask questions that can be answered with Yes or No or with other short responses. Where problems are obvious, such as in communicating an oral report, provide for an alternative channel. These might include written reports read by another student or the use of an interpreter. Be sure students with speech impairments feel comfortable in expressing a need by whatever method works best.

Alternative Strategy

A speech difficulty may be physiological, psychological, or some combination of the two. For problems of a physiological nature, consider technological alternatives, such as communication boards. Be sensitive to social or environmental factors that interfere with the student's speech, and do what you can to reduce the barrier.

Case Example

Luke is a severe stutterer. While his speech is intelligible for the most part, the stuttering pattern is more pronounced when Luke speaks to classmates with whom he has not previously been teamed in cooperative-learning tasks. Noting this, the teacher begins rotating Luke through various teams until he feels more comfortable. By example—slowing down her own speech rate—she guides all the students in being patient when Luke talks.

Emotional Disorder

Students with emotional difficulties may choose inappropriate ways of expressing needs, such as outbursts or withdrawal. The student with this type of disability should realize that these avenues are less productive than open, reasonable communication. Phase in appropriate requests for help and phase out ones that are inappropriate by using consistent, patterned interaction coupled with positive reinforcement.

Alternative Strategy

With the student, work out a contingency plan in which inappropriate forms of response are replaced by appropriate forms. Develop a corresponding schedule for applying positive reinforcement.

Case Example

Gilberto is bright but covers up feelings of inadequacy with overly aggressive behaviors. In a unit of study involving magnetism, the teacher has students make lists of things a magnet will attract. While they are doing this, the teacher looks at Gilberto's list and sees that a number of his items are not going to be attracted. Rather than set up a situation in which Gilberto is publicly proved wrong when lists are shared with other students, the teacher invites him to tell her his decision-making process as he thinks about each item. Together, they work through a few correct and incorrect items on the list. After it becomes clear that he is applying the right criteria in reaching his decision about what will and won't be attracted, the teacher asks Gilberto to review the rest of his list and make further changes and additions.

GATHERING INFORMATION

hearing impairment
physical impairment
vision impairment
learning disability
speech impairment
emotional disorder

As students become directly engaged in doing science, they should be aware of the importance of doing it right. Doing science right means several things, but chiefly that

- it is done safely, with minimum harm to living things or the environment

- a purposeful plan is followed

- accurate information is gathered during the activity

- the information is compiled in a useful, meaningful way

Students with disabilities will especially benefit from the orderliness of this approach.

The principal question to be resolved in information gathering is *how* the student can do it when the disability seems to interfere or the tasks apparently exceed the capacity of the student. Each case will be different, of course, but two adaptive approaches will be common—the use of teams, with a concomitant division of labor, and the use of technology to help in the accomplishment of the task.

Hearing Impairment K–12

Hearing-impaired students may have difficulty participating in group discussion. The teacher can facilitate hands-on activities and the collection of information by demonstrating or putting the steps in writing and giving examples of the kinds of information to be collected.

Encourage the student to make his or her own mind up about the steps and the kinds of information that should be collected. The student with a hearing impairment will naturally emphasize non-audible channels for information collection. Such a self-check approach calls on the student to ask, "Will I miss important information by ignoring auditory channels?" Encourage the student to develop techniques for acquiring the missing auditory information through a hearing classmate.

Alternative Strategy

Think through the information-gathering task that the hearing-impaired student faces. If it involves auditory information, consider alternative strategies together. If appropriate, use a buddy system.

Case Example

Ryan, who is profoundly deaf, is part of a group studying weather. The group is trying to determine why lightning and thunder are not perceived at the same time. From this they will arrive at a comparison of the speed of sound and the speed of light. They will then design and conduct an experiment to demonstrate the difference. During the next thunderstorm, each team member is to count the seconds between the flash and the boom. They will compare their counts. Ryan cannot hear the thunder, but at the teacher's suggestion, he touches a window to feel the vibration caused by the thunder. He collects and compares his information with that of the other students.

Physical Impairment

By definition, gathering information calls for active involvement. Because students with physical impairments may have problems in reaching, grasping, and manipulating, thought should be given to whether they *can* collect the information unaided. Is the information directly accessible? Can the specimen or equipment be moved? Is a partner needed? Is a special tool or mechanism required?

Examine each situation and disability in terms of the feasibility of gathering essential information. Having the student participate is the major objective; however, safety must also be considered. If safety may be compromised, try to involve a volunteer to minimize the risk (e.g., act as remote hands or legs) while carrying out the task at the direction of the student.

Alternative Strategy

Anticipate areas of difficulty and involve the student with the disability in doing the same. Together, work out substitute procedures while trying not to disengage the student from the activity.

Case Example

A field trip is planned to a local stream. Among the research questions to be answered is why the bank washes away faster on the outside of a curve. Tests are designed 1) to collect dirt from samples from each side to see if they differ in composition, 2) to see if the water on one side is deeper than on the other side, and 3) to see if the water moves faster on one side than the other.

Katrina uses a wheelchair. For test 1, another student collects the samples while Katrina directs. For 2, Katrina makes the observation visually, perhaps noting the appearance of lesser or greater depth in the water. For 3, a small water wheel is constructed using a spool and paddles (one of which Katrina paints white), loosely attached by a nail to the end of a broom handle. The site is selected so Katrina can reach the water by holding the broom handle from each bank. As the paddles spin, she counts the white paddle appearances aloud; another student times the experiment for 60 seconds.

Vision Impairment

Information gathering in science is highly dependent on observation skills. A large part of that observation, though not all, involves seeing various phenomena, making visual discriminations, and watching for cause and effect. Because of this, information gathering represents a special challenge for persons with visual impairment. It is important to be sensitive to the differences in order to take appropriate action on the students' behalf.

The nature and extent of the impairment will affect the information-gathering process. Low vision, for example, does not preclude the direct observation of some processes and specimens if the student has access to image-enlarging or image-enhancing devices. Similarly, using computer technology, blind students can extract information from printed source materials. Nevertheless, for both blind and low-vision students, a key to participation in data gathering remains the use of sighted partners who can orally report on what they see as experiments are conducted.

Alternative Strategy

Meet with the special education teacher to determine the extent of the student's vision loss. Clarify the availability of any devices that can potentially help the student extend his or her process of observation (consider school and community sources of devices). Plan experiments that lend themselves to two-person or small-group investigation so team members can share in the gathering of vision-based information.

Case Example

Ana is legally blind. In a science project involving meal worm behavior, a shoe box top is used as the containing area for staging the investigations. By using a CCTV image magnifier, Ana can directly observe the effects of different tests she and her partner perform.*

---

*See Telesensory Systems in the Resources chapter (p. 140).

Learning Disability

For most information gathering, students with learning disabilities will function within the range of skills exhibited by students who have no disability. One area of difficulty may be in the making of accurate observations and the accurate recording of data. Minimize this susceptibility to mixing up information by having students make multiple checks of the information either by repeating the investigation several times (an appropriate scientific procedure anyway) or by having partners collect the same information to verify it.

Alternative Strategy

Encourage the student with a learning disability to emphasize accuracy in experiments by being patient while observing details, checking data by means of repeated trials, and reexamining recorded information to be sure it matches what really took place. While accuracy is to be encouraged for all students, the learning-disabled student may need to check more carefully or often.

Case Example

August has a numerical-processing disability. In running some timed experiments with chickens, he and his partner take turns counting the number of pecks per minute. One counts pecks and the other watches the time. They switch roles several times to be sure that the recorded amounts are within a credible range.

STUDENT-CENTERED ACTIVITY

OBTAINING AND USING MATERIALS, EQUIPMENT, AND SPECIMENS

hearing impairment
physical impairment
vision impairment
learning disability
speech impairment
emotional disorder

Teachers sometimes encourage students to organize and set up experiments from the first step so they develop a sense of self-reliance and feel that the investigation is truly theirs. They not only design the investigation, but also gather the materials, equipment, supplies, and specimens. Depending on the project, students with some types of disabilities can find this process as difficult as conducting the investigation itself.

If the materials, equipment, and supplies are all in the classroom, problems are likely to be minimal, though still present. If collecting specimens is involved, difficulties may increase. For the most part, difficulties will occur in these areas: gaining physical access to needed items and their manipulation, seeing the objects to be obtained and used, and discriminating (carefully) the contents of materials containers.

Physical Impairment

To get a full feeling of involvement and responsibility, students with physical impairments should participate, as much as possible, in collecting and setting up the materials, equipment, and specimens that are to be a part of a science project. Clearly, lack of total accessibility will limit participation to some extent, as will each individual's limited capacity to arrange, construct, and manipulate the items assembled. For example, unscrewing a cap jar or tightening a clamp are dependent on muscle strength and coordination. Reaching for things is another area of potential difficulty. In advance, consider the requirements of a project and take steps to minimize difficulties that can be anticipated.

Alternative Strategy

Early in the school term, establish what items (materials, equipment, specimens) can be reached by a student with a physical impairment. If alternative arrangements are not possible, designate a student to obtain the items for the physically impaired student. For specimen gathering, use teams of students.

Case Example

Angie, a high school student, walks with the aid of crutches. Before class experiments begin, a science center lab assistant places the needed items on a table where Angie can reach them. For collecting specimens (e.g., garden insects), Angie is partnered with another student.

Vision Impairment 9–12

Students with vision impairments can locate and manipulate objects that are a) consistently kept in the same place; b) distinguishable by tactile means; and c) if all the same, such as jars or boxes, distinguished by tactile labeling. Collecting specimens can present a practical problem. If the collection process involves hunting for small objects, as in digging up worms, the task is probably not worth attempting unless a partner shares in the activity.

Alternative Strategy

Decide how to label supplies that cannot be distinguished by touch. For specimen-collecting activities, use partners unless the collection is done in a small space where specimens are likely to be encountered. Alert the blind student in advance to the kinds of tactile experiences he or she will be likely to encounter for the first time.

Case Example

Francine is blind. In a project on animal behavior, she and a friend collect garden snails. They also collect a variety of natural materials (leaves, stones, dirt) and manufactured materials (cans, foods, household chemicals) in order to conduct studies on avoidance behavior. Francine uses Braille labels to distinguish between her food and chemical jars.

Student-Centered Activity

USING COMPUTERS

hearing impairment

physical impairment

vision impairment

learning disability

speech impairment

emotional disorder

Computers have become an integral part of research in the physical sciences, life sciences, and earth and space sciences. To appreciate what scientists do with computers, and perhaps begin preparing for possible careers in the sciences, students should have the opportunity to do hands-on work with a computer.

Fortunately, the characteristics of computers often encourage their use by persons with disabilities. Students with learning disabilities are apt to write more clearly and revise more willingly when they can use a computer. Hearing-impaired students can function relatively easily in computer environments if audible signals are transformed into visual signals. Recent advances in software and peripheral equipment make computer use practical for people who are blind and vision impaired and, for advanced work, computers are preferable to manual alternatives. Similarly, special input controls have been designed to permit computer use by persons whose muscle control is so lacking that nearly any other purposive motor activity is impractical. Rapid advances in computer technology have led to a range of types of equipment, including portable equipment and peripherals that allow projection of the computer screen via overhead projection.

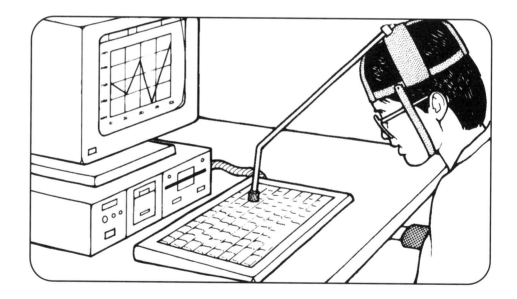

Physical Impairment

Students with physical disabilities differ widely in terms of their physical capacities. We are reminded by the media, and witnessing it first hand, that a physical disability can be compensated for in a number of ways. Film and television have shown a double-amputee basketball player, a wheelchair-using winner of the Boston marathon, an armless woman who drives, shops, cooks, and writes checks, and a paraplegic climber scaling a sheer stone mountain at Yosemite National Park.

Many students with physical disabilities welcome the use of a computer in learning because, in addition to the element of technological appeal, the computer does not demand physical exertion or the use of lower limbs, yet provides active engagement. If you have not previously used computers in the teaching of science, the presence of the physically disabled student in the class may be a good reason to do so. Involve the school district's technology staff when arranging any necessary adaptations. Rehabilitation organizations and large and small computer firms have concentrated on adapting computer technology so that it can be used by persons with limited use of arms and hands.*

Alternative Strategy For students who cannot use the computer because of physical limitations in their upper body, explore avenues for obtaining adaptive software, keyboards, special switches, and other special equipment. Contact vendors directly or go through school district technology specialists, special education teachers, or rehabilitation specialists in the community.

Case Example Elliot has muscular dystrophy and cannot use a regular keyboard. School district technology staff are consulted and in turn talk with a specialist in the computer company that supplied most of the district's computer equipment. Contact is also made with a rehabilitation specialist and with a manufacturer of special control switches. Soon a special keyboarding aid is installed in the science lab.

---

*See, for example, Articulate Systems and ComputAbility in the Resources chapter (p. 138).

Vision Impairment 3–12

Although it is not necessarily inexpensive, special equipment is now available (for example, from IBM and Apple) that facilitates computer use for the blind. School districts and departments of rehabilitation can work together to obtain equipment adapted to the individual or to a group of blind students who will share the equipment.

For students with low vision, recent computer models from major manufacturers already have a built-in capacity to greatly enlarge screen images.*

If adaptive computer technology is not currently available in your school district, have blind or vision-impaired students work as part of a team. Encourage sighted members of the team to vocalize about what is on the screen and what keyboard actions they are taking.

Alternative Strategy

If the classroom lacks special adaptive equipment, use a cooperative-learning strategy. Alternatively, arrange for an older, sighted student to volunteer to work one-on-one with the vision-impaired student at the computer.

Case Example

Ilsa's vision is such that she can make out only blurred forms and colors. The school has not equipped her or the science lab with Braille output equipment. However, on advice of a rehabilitation specialist, they purchase a speech output device that reads screen content. Ilsa soon becomes a star computer user.

The class undertakes a project of sorting and classifying seashells by characteristics. Several teams begin measuring and collecting classification data, which are then stored as a database in a spreadsheet format. Ilsa helps sort the seashells, using shape and texture variables, and enters her team's data in the computer.

---

*See, for example, Berkeley Systems in the Resources chapter (p.138).

Learning Disability **3–12**

Many learning-disabled students are eager to use computers. They are more than willing to do the same school work on the computer that they try to avoid in a paper-and-pencil format. While this is a positive argument for encouraging them to use computers, teachers should not leap to the conclusion that the individual's perceptual or cognitive problems no longer exist once the computer is turned on.

Variations in software can make a tremendous difference in what educational value the learning-disabled student derives from computer use. Apply common sense in your choice of software for a student with this learning disability. In particular, avoid a) cluttered screens with a busy look; b) distracting themes such as war-oriented "instructional" materials; c) densely spaced fonts (small size, single spaced) that fill the screen; d) sounds in the program that are disruptive; and e) speeded exercises involving eye-hand coordination that are not adjustable to each student's optimal rate. Word-processing software that is simple and has spell-check capability can be excellent for report writing.

Alternative Strategy   Select software that is not confusing in appearance and use. If complexity cannot be avoided, have student teams do the computer work.

Case Example   Tracy has a math-related learning disability and has difficulty solving word problems in science. As part of a cooperative-learning team, Tracy is involved in a computer simulation program called *SimAnt.** Through this software, the team learns that the control of variables has a direct effect on the welfare of the ant population. Tracy's attentiveness to math accuracy increases as she sees the practical effects of her decisions.

---

*See Maxi Aids in the Resources chapter (p. 139).

Speech Impairment 3–12

Students with problems in auditory reception of speech frequently find computer text to be a welcome alternative. Many speech-impaired students consider the computer and word processing as a happy alternative to vocalization. Moreover, speech output devices are now available that can facilitate oral (voice output) reporting, generated from the prepared written material stored in the computer. If possible, allow speech-impaired students to use this form of project reporting.

Alternative Strategy Provide a choice of reporting formats for science projects. At the same time, through informal socialization strategies, try to stimulate and encourage speech practice in the classroom as a way of building self-confidence in coping with the disability.

Case Example Chris is in the early elementary grades and is very sensitive about his poor speech, which has led to his being teased by other children. Chris is interested in science and, with his father, an engineer, he has learned to use a computer.

The science teacher accepts his written reports, in lieu of oral reports, but still calls on him to participate in classwork. To facilitate his participation without replacing it, the school district agrees to acquire a speech output device for the science center computer.

STUDENT-CENTERED ACTIVITY

Section B

Student-Centered Activity

hearing impairment

physical impairment

vision impairment

learning disability

speech impairment

emotional disorder

RESPONDING TO OTHERS AND
REPORTING FINDINGS

As science instruction moves away from lecture formats and becomes more discussion and activity centered, students can no longer remain passive. Participation and interaction call for active responding—raising questions, making contributions, exchanging ideas, reporting observed effects.

Students with disabilities may be reluctant to speak up and join in. This may be due in part to real barriers of speech impairment, but it also is frequently a reflection of lower self-confidence, often born of past experience and the expectation that they will get less positive feedback even if they do make the effort to respond actively. With this in mind, provide appropriate opportunities and ample positive support. Students with disabilities have the right to expect a receptive environment at school. Pay attention to their behavior and your handling of that behavior. Do you encourage or discourage active responses and reporting?

Hearing Impairment K–12

Hearing-impaired students' responses can be affected by a) whether they can see the person who is speaking, (b) the presence of competing stimuli, c) their vocalization skills, and d) acceptance by the teacher and other students. In mainstream classes, where interpreters are likely to be available in limited numbers, hearing-impaired students may find it difficult to follow the discussion, feel lost, and therefore be unwilling to participate actively.

To reduce confusion and involve the hearing-impaired students so they feel they belong, focus on one topic, with one speaker, at a time. Alert the hearing-impaired students to an aspect of the activity to which they can easily contribute. Call on them early enough so their contribution will not be a duplication of something another person already said (unless duplication is acceptable in the context).

Alternative Strategy

Position hearing-impaired students so that they can see (hear) each discussion and so that their better ear is toward the teacher. Cue each such student to contribute in such a way that other students will attend and be informed.

Case Example

Dee is hearing impaired. She can hear students and the teacher only if they are fairly close to her and speak one at a time. In a project planning session that requires brainstorming (on how to measure the weight of objects too small for a scale), the teacher has Dee's group meet in a corner of the room away from the other planning groups. Starting with Dee and moving clockwise in the groups, each person gets a chance to give his or her idea about how it can be done. Dee's speech is intelligible but halting. She uses a chart to help in her presentation. The teacher reinforces her participation and restates her idea for the others before proceeding to other planning groups.

STUDENT-CENTERED ACTIVITY

Physical Impairment

Generally, physically impaired students will be able to interact freely with their classmates and will be able to report the outcomes of experiments they have carried out. Perhaps one area of difficulty that can be anticipated is when students are expected to respond by using the chalkboard. In such a situation, be prepared to offer wheelchair users and students of short stature an alternative way of participating.

Alternative Strategy

Make contingency plans for any activity that calls for responses inconsistent with the physical capacity of the student. Insofar as possible, use alternatives that approximate the experience that other students have. For example, lap white boards or large pads of paper can be used in lieu of the regular chalkboard.

Case Example

Albert has paralysis of the lower extremities. Following a lesson on energy conversion in which teams have researched different forms of conversion, the teacher asks three students to go to the chalkboard along with Albert, who takes a position at the overhead projector, in a four-way "face off." The face off is a fun contest in which each team leader lists useful energy conversions. The team helps the team leader—in this case, Albert—by sending him notes with ideas of their own.

Vision Impairment

Vision-impaired and blind students do not experience difficulty in responding to the teacher or to others except when they must respond to a strictly visual event (e.g., recognizing identifying characteristics in an object held aloft by the teacher) or in an instance that requires eye-hand coordination (e.g. sketching or labeling specimens). Use cooperative-learning teams to help circumvent these difficulties.

Alternative Strategy

Be aware of the feasibility of calling for certain kinds of responses from vision-impaired students. If an inappropriate response cannot readily be avoided, pair the students with nondisabled students who can make appropriate responses on their behalf. However, call on the vision-impaired students as often as other students to keep them involved.

Case Example

Katerina is legally blind. Class activity is a culminating lesson on growth of fungi. Students have been growing specimens with different food bases in different environments. The class has reached a point in the lesson where they are also supposed to describe the results of their projects. Katerina's samples are as good as any of the others in class, but she cannot see them well enough to describe them. Each day the teacher has Katerina consult with a classmate, who gives her a verbal description while Katerina takes notes or puts descriptions on audio-tapes. The summary of these notes, along with her specimens, gives Katerina a basis for her oral report.

Learning Disability

Learning-disabled students can both respond and present findings. Keep in mind, however, that some of these students have an auditory-based disability. What they think they have heard may differ from what you said, so repeat the question to be sure it is understood.

In contrast, some learning-disabled students have adequate auditory perception but have difficulty in organizing what they know and presenting it clearly. Such students may project confusion in their presentation even though their comprehension is acceptable. Gently question the student to find out what he or she really knows.

Alternative Strategy   Discuss each student's learning disability with the special education teacher to know what instructional steps to take. Routinely allow all learning-disabled students extra time for responses and the preparation and delivery of reports.

Case Example   Jerome has problems with short-term recall of new vocabulary. In a science unit on weather, he learns many new words (e.g., *relative humidity, atmospheric pressure*). Knowing that class discussions will follow in which these and other words will be needed, the teacher has Jerome prepare a tip sheet that gives each word and a graphic reminder. She lets him refer to it during class discussions.

Next to the words *relative humidity,* Jerome draws water drops and labels the sketch "60%" to remind him about the units of measure for the word. Next to the words *atmospheric pressure,* he draws a balloon with arrows facing both out and in. Beneath the balloon he draws a peaked, wavy line representing the surface of the sea, to remind him that measurement is relative to sea level.

Speech Impairment K–12

Although interaction may be slowed by an inability to speak or by difficulty in speaking intelligibly, nonverbal individuals can communicate effectively by using a communication device. Many variations exist; some are simple communication boards displaying letters and words that the individual points to in some manner, while others involve electronics and some form of visual display.

Although use of such a device may slow class interaction at first, experience speeds the process considerably as the teacher and the student's classmates become adept in anticipating phrases that the speech-impaired student is generating.

Alternative Strategy

When a student's speech cannot be understood or a student cannot speak, explore the possibility of acquiring a communication device through the special education teacher, rehabilitation sources, or through a social service organization. Acquaint the other students in class with how the device works, perhaps even letting them try it.

Case Example

Tillie has cerebral palsy and cannot speak. Francine is deaf and speaks unintelligibly. Both are in the same science class and both are high achievers. They are involved in group activities aimed at demonstrating that algae produce oxygen and protists use oxygen.

In communicating her ideas and reporting on her group's work with algae, Tillie uses a non-electronic communication board. She points out the letters and words with a head-worn wand. Francine, in reporting her group's work with protists, uses a TDD device. She rapidly types her words on the keyboard, using a number of abbreviations, and they are instantly displayed on the small screen and can be read aloud to the group.

Emotional Disorder

The student with an emotional disorder who functions in a mainstream science setting can respond and actively participate in class activities. Nevertheless, the inclination to do so cannot be taken for granted. It is just as bad to have a disruptive student who uncontrollably breaks in on others' conversations as it is to have a student who is introverted and only participates when no alternative is avoidable. Creating a comfort zone for the student can be helpful.

Talk with the special education teacher to identify areas of need (e.g., tendency to be withdrawn, tendency to become aggressive as a defensive mechanism) and to map out steps to avoid situations that may trigger these behaviors. In particular, recognize the stress level associated with different forms of response in the classroom.

Early on, involve the student in a small-group project with classmates who are gregarious and not given to provocation. Try to avoid situations that will lead to failure and embarrassment. As the student's comfort level rises and when a safe topic is available, encourage the emotionally disordered student to be a spokesperson for the group, reporting to the class as a whole.

Alternative Strategy

Call for responses and participation commensurate with the student's socialization skills. Gradually increase the challenges in the student's participation while providing increased positive reinforcement.

Case Example

Roberto flares up when things go wrong. His aggressive manner is never on display when he is working with science projects per se, but his social skills are lacking. To engage him productively in group work, the science teacher puts him together with three other students who are social and friendly.

The group is to develop a food pyramid for a carnivore and an omnivore, relating the ultimate energy source to the sun. They begin in a library, where quiet study is the rule, and later meet to compare notes and develop charts. Roberto is invited to present the group work to the class as a whole.

Section C
Partners and Team Cooperation

hearing impairment

physical impairment

vision impairment

learning disability

speech impairment

emotional disorder

ORGANIZING PARTNERS AND TEAMS

Emphasis has been given to having students learn science in cooperative educational settings. A fundamental assumption of *Science Success* is that students will be organized in small groups in which science activities and projects are the shared responsibility of all. How can the science teacher structure these teams so students with disabilities and students without disabilities are comfortable and work effectively together?

The first thing to remember is that your feelings will be reflected by the students. If you feel that a student with a disability is not going to fit in, the other students will probably respond the same way. Neutralize the tendency to assign student pairings based on popularity. Instead, use a blind draw where none of the partners get to choose each other and team numbers are picked from a box. Later, after the relative skills and patterns of support and teamwork emerge, you can make purposeful match-ups.

Among the most important aspects of cooperative-group learning are the social interactions that occur. For reasons that vary from one case to the next, some students with disabilities may not have developed effective social interaction skills. Such students will need help in this area. You may not be an expert in social skills training, so you should ask for help from the special education teacher and then follow the intervention plan that you develop jointly.

Pay close attention to how tasks are divided within the team. Give the student with a disability various roles, and arrange for a back-up on the team if barriers are likely to be encountered.

Hearing Impairment

Hearing-impaired students are extremely conscious of the fact that much communication goes on around them but they can only participate in a bit of it. Combating this isolation is a critical element in making hearing-impaired students feel they belong in the regular science class. Using cooperative teams is a good way to break down the isolation.

At first, try to make pairings without favoring a student with a hearing impairment. However, if one or more students who have learned to sign or otherwise have developed a knack for communicating with the hearing-impaired student, it is not a bad idea to have them be part of the same cooperative group. Later, be sure that more and more students have the opportunity to work with the hearing-impaired student.

Alternative Strategy Try to give the student with a hearing impairment a meaningful role. Design the team so each student must directly address the hearing-impaired student in order to contribute his or her share. Avoid circumstances in which the team can work around the student, further contributing to isolation.

Case Example Candy is deaf but otherwise energetic and interested. She is placed in charge of a team that is trying to identify the various animals that live in the schoolyard soil and play a part in the ecosystem. Each student is given a task and reports to Candy. She has two students help her when the time comes for the class presentation.

Physical Impairment `4–12`

In structuring teams in which a student with a physical impairment participates, make sure all team members must cooperate to get the job done. Where one student has a problem, the others are expected to fill in the gap. Tell each team they will not be evaluated by how fast or how strong they are but rather how well they use scientific methods to produce good answers to meaningful questions.

Alternative Strategy

Rotate student assignments to teams in order to have all students share in the accomplishment of team goals. Within teams, use a back-up model so that any difficulty encountered can be cooperatively resolved.

Case Example

A class project involves using various instruments to create sounds and then studying wave forms and frequency. Geoffrey, who is hearing impaired, is assigned to the string instruments because he can feel the strings vibrate and differentiate high and low frequencies. Geoffrey borrows his uncle's audio player, which converts frequencies into different colors, and explains to the class the principle on which it works.

Vision Impairment 4–12

Guard against letting teams function in a way that bypasses the student with a vision impairment. Try to ensure that each student in cooperative groups gets to do a wide variety of things. That is, shift roles from time to time. When a mismatch arises and vision is central to a task, have the vision-impaired student select who will serve as his or her eyes for the task, then select different students for other tasks.

Be particularly aware of the tendency of nondisabled students to make decisions for the student with a vision impairment. Loss of vision does not reduce a person's ability to think. It may mean that the vision-impaired student needs to have someone be "temporary eyes" so that he or she can get the information needed for making good decisions. Help partners serve as communicators about the environment, and help the student with the vision impairment by being sure the task or project includes elements he or she can do.

Alternative Strategy As cooperative groups begin their work on an activity, set aside an initial block of time for them to plan ahead. Have them decide what needs to be done and who will do it. Be certain that the roles and tasks include the vision-impaired student and that there are back-up persons.

Case Example Augie has severely limited vision but is able to use the computer when the image is greatly enlarged. In a cooperative team working with the computer, Augie takes his turn at the keyboard. The other students help him keep his place on the screen. Augie inputs answers to questions posed in the software.

Learning Disability **4–12**

Students with learning disabilities may be prone to mistakes or slow in completing tasks. Because of this, they may encounter resistance from other students when first assigned to a team. Be alert to any persistent problems of socialization that a learning-disabled student may have. Discuss these with the special education teacher and jointly plan a strategy for dealing with the problem.

Make it clear that team members are not in competition with each other but are trying to reach goals that they set together. The key to cooperation is to have students learn that they will do better by supporting each other than by competing. For example, the learning-disabled student may have a knack for conceptualizing problems that other students could learn from. Finally, so that the student with a learning disability is helped to the fullest extent, use a lot of positive reinforcement with nondisabled students who have played a constructive role in helping the learning-disabled student.

Alternative Strategy

Form teams randomly by picking numbers from a hat. Meet with the groups to which the learning-disabled student belongs and help them set goals and plans. Such plans may include the use of backups who help members reach the groups' goals.

Case Example

Alex has both a math disability and attention disorder. In a team project on weather, Mary is Alex's backup to double check the temperatures he records each day. To keep him attentive and involved, the teacher chats with his group and has Alex tell her what each person is working on.

Speech Impairment 4–12

Much of the activity in science is quiet activity, but not all. When speech is important and one or more students have some type of speech impairment, teamwork becomes even more important. People vary greatly in how well they can understand severely flawed speech. Students may need to be coached to learn this skill. They should be encouraged to be patient and see the speech-impaired student as a person, much like themselves, with much to offer. It is important that at least one student who understands severely flawed speech be a part of the team in which the speech-impaired student participates.

While help is important, so is the speech-impaired student's responsibility to try to generate speech that can adequately convey questions (as in an interview) or information (as in explaining a procedure). Peers on the team and in the class should respect the speech-impaired student's skills and be patient in hearing him or her out when speech is necessary.

Alternative Strategy

Students take turns being the speech backup for the speech-impaired student. Start with a student who does this well and who is a class leader. Involve the speech-impaired student in practice communication with members of the cooperative group to help them work together.

Case Example

Michelle speaks haltingly and with a spastic slur. She is bright and knows answers as quickly as anyone. For projects she is teamed with a backup student who knows her well and can "interpret" what she says.

Emotional Disorder 4–12

Students with an emotional disorder may play that disability out in ways that are disruptive to teams of peers. They may seek attention in inappropriate ways or resist or evade it. In either case, the nondisabled students should not be penalized when such a student is part of their team. However, it is both helpful and practical for them to be supportive of this student, just as they would help any other student.

To be helpful, students must have a good example in cooperative learning from the science teacher. While lots of positive reinforcement is the preferred approach, peers may get the best results by simple phrases like "you can do well when you give it your best try."

Alternative Strategy

Use a plus and minus system, giving pluses to the team when team members work constructively together, and a minus if no effort is made or disruption has taken place. Be liberal with pluses.

Case Example

Buster has been diagnosed as having an emotional disorder. He is also hyperactive and seeks attention whenever he is on unsure ground. To control this, the science teacher uses two techniques: she institutes a time-out procedure whenever Buster becomes disruptive and must be removed from the group activity, and she credits the whole group for any day in which no disruptions or problems occur in the group.

Partners and Team Cooperation

hearing impairment

physical impairment

vision impairment

learning disability

speech impairment

emotional disorder

SHARING TECHNOLOGY

The use of technology in science classes is becoming more and more common. Not only are students becoming computer-literate at earlier ages, but they are no longer dazzled by visual media and audio materials. This sophistication, however, may not hold true for the student with a disability whose prior experience may be more limited. Care needs to be taken to assure that this student can learn to use the technology along with his or her peers. At the same time, special-purpose technology, such as a sip-and-puff computer-controlled mechanism, can provide a learning experience to other students even though it is acquired principally for use by the student with a disability.

Science teachers are often comfortable with technology and are good at imagining how switches, clamps, and control devices (see Chapter 5, Resources) can be adapted for special-purpose use in science centers or classrooms. Imagination has reduced many barriers through clever use of technology. Chapter 5 suggests some of the many manufacturers and vendors of technology for persons with disabilities. Literature from IBM (*Increasing Access to the World Around You*) and Apple (*Toward Independence: A Guide to the World of Macintosh Computing for People with Disabilities*) are good places to start your search for information about current technology alternatives.

Hearing Impairment K–12

Technology devices are becoming increasingly available to students with hearing impairments. One such device, designed to be shared by hearing and nonhearing individuals, is the *Lightwriter*.* This and similar lightweight, portable units allow typed messages to be displayed electronically in print or speech form. Using these devices, hearing and nonhearing students (and the teacher) can communicate with each other freely, without regard for lip-reading or speech and hearing impairments, and even via telephone.

The Television Decoder Circuitry Act of 1990 mandated that by 1993 all newly manufactured TV monitors must be able to receive closed captions.

Alternative Strategy

Deaf and hearing students can communicate with each other without relying on sign language. Review the technological alternatives with school staff and, once these technologies are acquired, introduce them as tools and as an example of applied science. Let other students try the devices if it is practical to do so.

Case Example

Petra is severely hearing impaired and speech impaired. She uses a *Talking Notebook II* to write class notes and draft reports.** She edits and displays her work on a high-contrast screen and activates her scratchpad to "speak" what she has written. With this technology, she can keep up with her classmates in science.

---

*See Tech Aid in the Resources chapter (p. 140).
**See ZYGO Industries in the Resources chapter (p. 141).

Physical Impairment $\quad$ K–12

Computer technology is very important for many persons with disabilities, including those who have trouble using a keyboard. Fortunately, a variety of recent technological advances makes this wider computer use possible.

Rehabilitation specialists usually know companies that customize computer controls to the needs of the individual, even to the extent of using sip-and-puff controlled technology (breath-controlled through tubes), tongue-driven, or eye-controlled systems. Another approach is the use of voice-recognition technology.* For those with limited muscular control, a number of adapted keyboards and key guards are also available. Robotic manipulators allow the severely physically impaired person to manipulate objects.** Special input systems for computers do not make them harder to use for students who are not physically impaired.

Alternative Strategy $\quad$ Adaptations that simplify access to computers will help the science-minded student with a physical impairment. Explore local and national resources to identify alternatives to meet the student's needs.

Case Example $\quad$ Rick has muscular dystrophy and severely limited motor skills. He has received an evaluation by a rehabilitation specialist who has suggested a voice-controlled input device that can be used with computers already owned by the school's science program.

---

*Selected voice-control devices and other special-purpose software are mentioned in the Resources chapter.

**See Regenesis Development in the Resources chapter (p. 140).

Vision Impairment   K–12

Computer use by blind and vision-impaired persons was once an impossibility. Now there are a number of choices for input and output, including both standard and highly portable technology. Among the choices are Braille output and speech output, and enlarging systems for screen displays and scanners to simplify input. Costs for these software and hardware alternatives range widely. In addition to computers, a variety of special-purpose technology is available from suppliers.*

Alternative Strategy

Obtain catalogues from suppliers of equipment for special-needs persons. Try to identify items that can benefit all the students in the class but especially the person with a disability.

Case Example

Felipe has low vision. The class is studying weather and uses topographical maps that Felipe can't see. From a supplier's catalog, the teacher orders a hands-free page magnifier for under $12 that covers a 7-by-10-inch area. Everyone in class benefits from sharing this equipment.

---

*See, for example, LS&S Group (p. 139), Maxi Aides (p. 139), Raised Dot Computing (p. 140), Telesensory Systems (p. 140), and the American Foundation for the Blind (p. 142) in the Resources chapter.

Speech Impairment K–12

Severely speech-impaired students share the problem of oral communication, but some are physically impaired in such a way that their written communication is limited as well. Because communication is important to any shared activities in which teams of students are engaged, disabled and nondisabled students benefit when speech-substitution devices are used.

Speech-substitution devices, or augmented communication devices, allow individuals to speak artificially through some variation of an electronic communication board (or keyboard). For those with many impairments, communication boards may involve either simpler strategies (such as head-band supported wands for touching keys) or more technically advanced strategies (such as devices operated by visual scanning).*

Alternative Strategy   Select a speech synthesizer or communication board based on an analysis of the student's needs (speech-only or speech-plus-physical) and his or her need for portability or access to computers.

Case Example   Marie cannot speak, but she is very active and involved in a variety of school affairs. In science class, she's part of a project in which intra-team communication, as well as communication in the community, is important. She does not yet need a computer but will in a year. She contacts Rehabilitation Services and with their participation selects the "Secretary" writing/talking communication aid because it is portable, allows up to 20 pre-recorded messages, and can be used for up to 52 memos or notes that can be printed on an internal printer-tape.**

---

*See, for example, ZYGO Industries (p. 141), Tech Aid (p. 140), and ComputAbility (p. 138) in the Resources chapter.
**See above note.

Section C

Partners and Team Cooperation

CARRYING OUT ASSIGNMENTS

hearing impairment

physical impairment

vision impairment

learning disability

speech impairment

emotional disorder

A central assumption about teamwork is that in order to maximize the performance of the team as a whole, it is important to integrate the complementary skills of the team members. This is true in sports, in business, and in science classes. As cooperative learning takes place, assignments should be designed to bring out the best in each team member while still allowing all members a chance to contribute in multiple ways.

Role taking, note taking, and data gathering are three important components of team learning. Inherent in all three is the general notion of turn taking, assuring that the student with a disability is given a fair chance to do hands-on work insofar as he or she can do all or some part of the activity. In making assignments, start with the skills that the student is able to do. As the team begins to function, gradually shift assignments so the student with a disability is expected to take on tasks that will involve some accommodation. Accommodation may take the form of a backup partner, the use of technology, or some shift from the usual way in which the science activity is done. Don't be afraid to ask the team to solve the problem and figure out the best way to share the work.

Hearing Impairment

Students with hearing impairments will be able to carry out most science assignments provided that the assignment a) is given in a form that they can understand, b) doesn't require sound discrimination beyond the students' ability, c) is not dependent on group discussions in which the student will be unaware of who is speaking, and d) the productive speech quality required is not beyond the skill level of the student.

Giving assignments is particularly problematic at the pre-reading grade levels where written materials cannot be substituted for speech effectively. At this level, the teacher should probably substitute demonstration and modeling by another student. Lip-reading may be marginally useful but it will not be effective if the child is young and many different speakers must be understood.

Alternative Strategy

Seat the student where he or she can easily see other students as they do their work or speak. Take time to explain or demonstrate what needs to be done.

Case Example

Sharilee is a second grader with a severe hearing impairment. She cannot hear sounds without amplification, and if the sound is coming from behind her, she cannot determine its origin. She is positioned so she can see both the teacher and the other students. The class is having a brainstorming session, and the teacher is trying to get the students' ideas about water rationing and why water control is necessary in drought conditions. After drawing a full glass, then a half glass, the teacher simulates getting no water from a faucet (pretending to turn the handle). Then she has a classmate help Sharilee make a drawing of a full dam and one that's nearly empty. She starts brainstorming with the classmates—clockwise, one at a time—so that Sharilee can see them as they talk and give ideas. When it is Sharilee's turn, the teacher shows her the dripping faucet and gets her to tighten it to stop the waste.

Physical Impairment 4–12

Apart from obvious mismatches between the disability and the physical demands in the particular assignment, students with physical impairments will be able to complete a wide variety of assignments. However, tasks that relate to data collection (gathering information by carrying out experiments) may be one area in which difficulties can arise. Try to avoid presenting *no* challenges to the student or insulating him or her from rich experiences and interactions with peers, but try to think of ways you and the other students can engage the student actively in the regular classwork.

In structuring teams in which a student with physical impairments participates, make sure all team members must cooperate to get the job done. Where one student (such as the student with the physical impairment) has a problem, the others are expected to fill in the gap. Tell each team they will not be evaluated by how fast or how strong they are but rather how well they use scientific methods to produce good answers to meaningful questions.

Alternative Strategy

Rotate student assignments to teams in order to have all students share in the accomplishment of team goals. Within teams, use a student backup model so any difficulty encountered can be cooperatively resolved.

If an activity cannot be accomplished by the student, consider a) a modification of equipment or procedures that are normally used, b) changing it from a one-person to a two-person activity, or c) designing a parallel but different experience that will lead to the same scientific outcome or understanding.

Case Example

Davis and his classmates are asked to design and carry out an investigation of factors that encourage or block the growth of microscopic organisms.

A short list of topics is presented to stimulate the students' choices of activities. These include

- testing effects of toothpaste on bacterial growth
- testing toxicity of common powders
- testing toxicity of household sprays

Fine-motor impairment prevents Davis from preparing slides and focusing the microscope, but he can look at slides that have been prepared by other students in the team. He can also prepare his own set of test samples in petri dishes with only minor help from a buddy.

Vision Impairment **4–12**

Students with vision impairments can make substantial contributions to groups. Although blind students will be limited in some tasks that are sight dependent, students with low vision can often see enough to participate on a nearly even footing. Where sight is essential, such as in creating graphics for a project report, another team member will have to take the lead.

Get to know the leadership potential of the students with vision impairment. They have often become pretty well organized in order to keep track of things, and they are often perceptive about people. Both attributes are characteristics of good leaders, so you might try giving them substantial amounts of responsibility within their team or partnership. They will have a pretty good idea of what they can and cannot do in carrying out the assignment, and given the opportunity, many will be able to coordinate the work of the team effectively.

Alternative Strategy

If feasible, add to the student's level of self-esteem by giving him or her a certain amount of leadership responsibility within the team or group. If the team assignment is such that sight is essential, you will need to identify the best way that the student with a vision impairment can contribute and what help will be needed from a partner.

Case Example

Casey and her classmates are studying how variables affect outcomes. Assignments are made in chemistry (chromatography), life science (animal adaptations to climate), and physics (ice cube melt test). Casey's vision prevents her from carrying out all three assignments independently, though she can do the ice cube assignment on her own. For the other two, Casey works with a partner, and the teacher helps her decide what roles she should take in each of these assignments.

Learning Disability

Although learning disabilities include a number of different, specific subtypes, many students with learning disabilities are more successful when they are able to engage the subject matter in a concrete, rather than abstract, form. Accordingly, they are more likely to encounter difficulties when the assignment is grounded in words, is complex and technical, or is abstract rather than directly observable, applied, and hands-on. Whenever possible, make class assignments concrete and observable.

Having noted this, be alert to special cases of learning-disabled students who do have excellent reasoning and critical thinking skills. For these students, the main problem may rest in perception of words or other symbolic cues.

Alternative Strategy

To the extent possible, supplement class assignments that are reading centered with activities that make the reading more realistic. Where the content is abstract, try to introduce simulations to make it more concrete. Use plenty of examples, oral or otherwise, to make topics more applied.

Case Example

Rhoda tends to be confused by complex printed matter. In a unit on nuclear waste and the environment, concepts are made clear to her by simulations in the classroom. A nuclear dump area is established within a certain radius of the site; students wear "protective" clothing and move one radioactive material without touching it. More radioactive waste is added, and the risk factors are revised. Students describe half-life and illustrate it with clock-time modifications. In all these hands-on activities, Rhoda plays a prominent role.

Speech Impairment

Communication is central to most cooperative-learning situations. For the speech-impaired student to benefit in such a setting, early decisions must be made about how the speech impairment will affect the group interaction. Similarly, strategies for facilitating interaction should be addressed as class assignments are given. Fortunately, most group assignments lend themselves to divisions of labor.

Initially, when the speech-impaired student is starting to work with a group, the tendency will be to talk around him or her and not to engage in direct person-to-person communication. This approach is not recommended. Rather, the emphasis should be put on *how*, not whether, the speech-impaired person can best make his or her ideas known in the group.

Alternative Strategy

Meet with the special education teacher to decide on how best to provide the speech-impaired student with a way to express thoughts or report information. For a few classes early on, work closely with the group in which the student is placed. Guide the nondisabled members in how to interact with the student (e.g., asking to repeat, waiting for the whole statement, repeating back to check on correctness).

Case Example

Dinah is bright but mute. Over several weeks, classwork will be spent on developing team positions and arguments on what are the top three contributions of science to society. Each team is to defend the ideas of its members, gather supporting evidence to identify the top three in the idea pool, and present oral arguments supported by published information.

In meeting with the groups as they begin, the science teacher stimulates thinking by collecting, in writing, up to five ideas from each person. She asks one vocal person to lead discussion as the group decides how to collect supporting information. Because Dinah has a TDD, the science teacher suggests that maybe she could use it to interview a scientist with disabilities to get his or her view. An interview is then conducted with a scientist named in one of several American Association for the Advancement of Science publications.* Each of the team members then takes a turn using the TDD to interview another scientist with a disability. Each team member also uses library resources to help prepare their group argument.

---

*See *Resource Directory of Scientists and Engineers with Disabilities* (p. 136) in the Resources chapter.

Emotional Disorder

As teams of students collaborate on carrying out assignments, there will inevitably be some disagreement, confusion, and frustration among team members. Regardless of the science topic under study, these conflicts can trigger the emotionally disordered student. He or she can become counterproductive to completing group-based assignments. To dampen these effects and discourage intra-group disputes, use positive reinforcement techniques, along with a schedule for reinforcement, to credit the group for *how* it works together and *what* its results are.

Alternative Strategy

Ask the special education teacher to help set up a reinforcement schedule for the student with an emotional disorder and to meet with the student to explain how it works. As it is initiated, do not focus exclusively on the student with a disability, but pay close attention to all students in the group and let them know they are doing a good job. Also, with the help of the special education teacher, devise a communication channel through which the student can express feelings without disrupting others in the group or the class as a whole.

Case Example

Aram is performing well academically, but he is prone to emotional outbreaks, apparently brought about when he internalizes his feelings and then cannot contain them. Aram raises exotic fish and shows an interest in science. To capitalize on his interest, and to minimize his emotional response pattern, he is made team leader for an assignment that relates to living things.

With the special education teacher present, it is explained to Aram that how well he leads the team (and shows self-restraint) and uses resources (requests intervention if needed) will add 50 percent to the outcomes of the science project. As the project begins, his team leadership is noted and all team members complimented.

A REVIEW OF KEY POINTS

This section has emphasized the variability of teaching students with disabilities in science settings. The different needs of these students arise as a consequence of

- different attitudinal or affective barriers these students encounter in dealing with others

- different levels of severity and types of disability that function as personal barriers

- different environmental barriers—particularly in the classroom, the science center, and field activities

Regardless of the facts that this variability exists and that each student presents different challenges, many barriers are minor. They can be directly addressed and overcome without undue difficulty. Generally, teachers will be most successful when they

- make an effort to expand their skills in adapting instruction to different students

- plan barrier-reduction strategies in advance with input from the student

- collaborate with the special education teacher, technology staff, and other school-district specialists

- use principles of hands-on learning, cooperative-team learning, and inquiry

- teach through problem solving that involves the application of scientific methods and the generalization of science concepts

- keep a sense of humor about each day's happenings

- remember that teachers serve as role models for showing respect for the interests and capacities of others and generating interest in science as a possible career field

Finally, it comes down to this: As a teacher you are already committed to helping young people develop and reach their full potential. You'll derive even greater satisfaction from your experience with students with disabilities. You will find it satisfying to see nondisabled students accepting and enjoying the participation of students with disabilities. You'll feel satisfaction as you observe them gaining knowledge and developing the competencies that will serve them well in adulthood. The interaction will be valued by you both. In a very real sense, as you teach them, so shall you learn from them.

Chapter 5

Resources

The resources listed in this section are divided into five topics:

- **Sourcebooks and Periodicals**—volumes and publications that can direct you to products and people to help students with disabilities (pp. 135–136)

- **Computer Database Sources**—assistance and information that can be accessed by computer (p. 137)

- **Adaptive Devices, Software, and Technology**—products to help persons with disabilities use computers or carry out other difficult tasks (pp. 138–141)

- **Disability-Oriented Organizations**—advocacy and assistance groups that can provide information about specific disabilities and available resources for persons with those disabilities (pp. 141–146)

- **Science and Technology Resource Organizations**—resources for science education in general (pp. 146–147)

The resource organizations and products are illustrative only. No endorsement of quality is implied by their inclusion; they illustrate the wide range of resources that can be tapped. We encourage schools to use their usual procedures for ensuring the quality and relevance of products.

SOURCEBOOKS AND PERIODICALS

Apple Computer Resources in Special Education and Rehabilitation (1988). Available from DLM Teaching Resources, One DLM Park, Allen, TX 75002, (800) 527-4747.

 This 400-page sourcebook contains information about more than 1,000 hardware and software products, publications, and organizations involved in helping children and adults with disabilities.

Assistive Technology Handbook (1990). Enders, A. and M. Hall, RESNA Press. Contact: RESNA Technical Assistance Project, 1101 Connecticut Ave., NW, Ste. 700, Washington, DC 20036, (202) 857-1140 or (202) 857-1199.

 This 556-page sourcebook for people with special needs includes chapters on information sources and consumer strategies for information seeking (databases, publications, information centers, research and development organizations, and clearinghouses), safety and consumer protection issues, environmental controls used at home, assistive technology at work, assistive devices (including toys) for young children, recreation technology, devices used for mobility purposes, communication and sensory aids, computer technology, control mechanisms, and emerging clinical and treatment methods (prosthetics and orthotics, functional neuromuscular stimulation, and other new procedures).

Case Management Resource Guide: 1992. Center for Consumer Healthcare Information, 4000 Birch St., Ste. 112, Newport Beach, CA 92660, (800) 627-2244.

This exhaustive resource contains detailed data on special services, credentials, and contact names for thousands of health care providers including associations in the disability area, clearinghouses and hotlines, independent living centers, and specific health care services or service providers in different counties. Four volumes divide the U.S. into four sections to focus on local resources.

Closing the Gap, P.O. Box 68, Henderson, MN 56044, (612) 248-3294.

This bimonthly newspaper publishes information about adaptive technology. A yearly resource directory lists vendors of hardware and software for disabled persons and organizations serving disabled computer users. Closing the Gap sponsors an annual conference on technology for persons with disabilities.

Increasing Access to the World Around You (1992). IBM Special Needs Systems, P.O. Box 13238, Boca Raton, FL 33429-1328.

This 20-page annotated catalog, with information sources, includes products for improving computer access from IBM and other manufacturers. Free.

Mainstream, 2973 Beech St., San Diego, CA 92102, (619) 234-3138 (voice and TDD).

This magazine for the able-disabled is full of information about current developments and includes a substantial number of advertisements for products of potential value in helping persons with disabilities become more independent.

Managing Information Resources for Accessibility (1991). Clearinghouse on Computer Accommodation of the Information Resources Management Service, U.S. General Services Administration. Contact: COCA, Rm. 2022, KGDO, 18th and F Sts., NW, Washington, DC 20405, (202) 501-4906 (voice) or (202) 501-2010 (TDD).

This comprehensive handbook is primarily a helpful tool for government agencies, but can be useful to any school or organization providing assistance to students with disabilities. Of special interest is Section III, "Overview of Accommodation Solutions," which has subsections for technology users with visual, hearing, or mobility impairments. The "Accommodation Resources and Information" appendix covers federal government resources and public and private sector resources, and the "Accommodation Products" appendix gives a representative list of products and suppliers. No charge for single copies.

Resource Directory of Scientists and Engineers with Disabilities (1987). Stern, V., D. Lifton, and S. Malcolm (Eds.) Publication 87-13. American Association for the Advancement of Science (AAAS), 1333 H St., NW, Washington, DC 20005.

This 139-page directory lists scientists and engineers with disabilities who have volunteered to be included by AAAS as advisors, collaborators, or consultants. Suggested uses for the directory are as a source of a) role models for families, teachers, and counselors to dispel the myth that science careers are not viable for persons with disabilities; b) information about how to apply coping strategies and technological solutions to scientific situations; and (c) speakers and advisors to community groups and others who are promoting greater participation in science by young people with disabilities. The list provides address and phone number, degree, areas of specialization, most recent position, nature of disability, and types of consulting interest.

Macintosh Disability Resources, Worldwide Disability Solutions Group, Apple Computer, Inc., 20525 Mariani Ave., Mailstop 2SE, Cupertino, CA 95014, (800) 732-3131.

This list of adaptive hardware and software for Macintosh computers can be obtained in either printed copy or Hypercard stack.

COMPUTER DATABASE SOURCES

ABLEDATA, National Rehabilitation Information Center, Ste. 935, 8455 Colesville Rd., Silver Spring, MD 20910, (800) 346-2742 (voice or TDD).

ABLEDATA is a large database of all adaptive technology products. Topical areas include products for Personal Care, Home Management, Vocational Management, Educational Management, Mobility, Seating, Transportation, Communication, Recreation, Walking, Sensory Disabilities, Orthotics, Prosthetics, and Therapeutic Aids. Single copies of fact sheets (e.g., tilt-in-space wheelchairs, stair lifts) are free. Custom searches of the database by staff, involving up to eight pages of product entries (approximately 24 products), are free; larger searches have nominal charges. Copies of the ABLEDATA thesaurus are available for a fee. The database is directly accessible through modem on a subscription basis from BRS Information Technologies, (800) 345-4BRS. Hypercard-controlled CD-ROM and floppy-based versions can be obtained from the Trace Center, University of Wisconsin, Madison, WI, (608) 262-6966.

Compuserve, America Online, and *GEnie* are among the national telecommunications networks with disability forums. Questions about disability issues and access can be addressed remotely through a computer modem.

Compuserve's Handicapped Users Database is a bulletin board with nine sections, including listings of computer networks, adaptations of computer hardware, software, and current research. Disabilities Forum is for on-line conferences on disability-related subjects. (800) 848-8199.

disABILITIES Information Services, 9840 Stanwin Ave., Arleta, CA 91331-5303, (818) 899-1598 (voice or TDD) or (818) 890-1130 (for data).

Information about this database service can be obtained by computer, telephone, TDD, fax, or electronic mail. A variety of information is available, including text of the Americans with Disabilities Act, agency lists, adapted programs, product reviews, message boards, support groups, and the opportunity to have real-time conferences with specialists in disability. The service is part of the GEnie Network.

National Center for Research in Vocational Education, TASSP Computerized Database, University of Illinois, Room 345, Education Building, 1310 S. Sixth St., Champaign, IL 61820, (217) 333-0807 (phone), (217) 244-5632 (fax).

An extensive, computerized database is available to conduct free information searches in the area of Technical Assistance for Special Populations Program (TASSP). Included are reference materials; research reports; monographs; journals; lists of organizations, associations, and agencies; centers for information services; clearinghouses; computer-based information networks; databases; contact persons in state and national agencies and organizations; and descriptions of exemplary practices identified by TASSP.

Project Enable, West Virginia University Research and Training Center, One Dunbar Plaza, Ste. E, Dunbar, WV 25064-3098, (304) 766-7842 (modem), (304) 766-7846 (fax). Contact: Information Systems Manager, (304) 766-7138.

Project Enable is a computerized information system with more than 1,000 files of text and software available for downloading. Information is available to both service providers and to persons with disabilities. Among the Enable discussion groups that can be directly accessed through use of a computer modem are a science education discussion group (K–12); a teacher discussion group; and elementary, middle, and secondary school "chats."

ADAPTIVE DEVICES, SOFTWARE, AND TECHNOLOGY

The main product lines of the companies are identified as being geared primarily, though not exclusively, toward the following sorts of students:

HI   Hearing impaired or deaf   **SI**   Speech impaired

LD   Learning disabled   **VI**   Vision impaired or blind

PI   Physically impaired

American Thermoform Corporation, 2311 Travers Ave., City of Commerce, CA 90040, (213) 723-9021. **VI**

This company is a leading supplier of materials and equipment used in the preparation of brailled materials. They also make the raised-line printers that make it possible for blind persons to feel the outlines of non-brailled shapes.

Apple Computer Inc., 20525 Mariani Ave., Cupertino, CA 95014, (800) 732-3131 or (408) 996-1010. Contact: Manager for Special Education and Rehabilitation. **HI, LD, PI, SI, VI**

Apple has joined with the Disabled Children's Computer Group, Berkeley, Ca., and the Alliance for Technology Access (ATA), in Albany, Ca., to promote a coalition of independent organizations dedicated to helping disabled children gain access to computers of all types, including the Macintosh computers that build in many adaptive features for disabled users. Currently over 40 ATA resource centers operate in the U.S.; they share information via Applelink. A list of the Centers is available from Apple or can be found on page 341 in the September 1990 issue of *Macworld* magazine. (See also, in the same issue, the article on adaptation entitled "With a Little Help from My Mac.")

Arkenstone Inc., 1185 Bordeaux Dr., Ste. D, Sunnyvale, CA 94089, (800) 444-4443. **LD, VI**

The *Arkenstone Reader II* consists of three components: a scanner recognition card, Arkenstone Reader software, and a scanner. Together, they scan printed text documents so that a number of computer applications can use the data directly. The company also sells *An Open Book,* a stand-alone reading unit.

Articulate Systems, Inc., 600 West Cummings Park, Ste. 4500, Woburn, MA 01801, (800) 443-7077, (617) 935-5656, or (617) 935-0490 (fax). **PI**

Voice Navigator combines software and hardware to allow voice control over the commands in an application. For example, with a Macintosh computer, *Voice Navigator* would allow the user to set formatting rules in a word-processing file instantly with voice directions instead of using a mouse and screen menus.

Berkeley Systems, Inc., 2095 Rose St., Berkeley, CA 94709, (800) 877-5535 or (510) 540-5535. **PI, VI**

Outspoken is a software product that can work on the Macintosh or other computer with a pixel-based display and a graphic interface. It uses the built-in Macintosh speech synthesizer and work with word processors, spreadsheets, and databases in place of or with the mouse. Berkeley also produces *inLARGE,* a screen-magnification program, and *Screenkeys* for use by the physically disabled.

ComputAbility, 40000 Grand River, Ste. 109, Novi, MI 48375, (800) 433-8872 or (313) 477-6720. **LD, PI, SI, VI**

An alternative input device for computers called *AID+ME* allows the user to control the computer through a membrane keyboard, joystick, scanner, switches, serial communication, or Morse code. It allows adjustments for repeating keys, multiple keys, and the length of time a key must be pressed to be read.

Don Johnston Developmental Equipment, Inc., P.O. Box 639, Wauconda, IL 60084, (800) 999-4660. **LD, PI, SI**

This company produces and sells computer access products such as the *Adaptive Firmware Card* and *KE:nx* for the Apple and Macintosh computers.

Fisher Scientific Co., 30 Water St., West Haven, CT 06516, (800) 766-7000 or (203) 934-5271. **PI**

This firm manufactures a portable lab station for physically disabled individuals and supplies general purpose scientific equipment.

International Business Machines Corp. (IBM), P.O. Box 2150, Atlanta, GA 30301-2150, (800) 426-2133 (voice) or (800) 284-9482 (TDD). **HI, LD, PI, SI, VI**

The Special Needs Information Referral Center is an information center on assistive devices, including IBM's products that facilitate the use of computers by persons with disabilities. Technical support staff can answer questions about IBM's Independence Series of Products, including the IBM *Screen Reader* for blind and vision-impaired readers, IBM *Speechviewer* for persons with speech or hearing impairments, *VoiceType*, IBM *PhoneCommunicator*, for persons with hearing and speech impairments, and *THINKable* cognitive rehabilitation tool for direct client therapy. IBM's *Disabilities Assistance Network* is a program that loans personal computers at no charge to selected state agencies and nonprofit organizations that receive federal grants under the Technology Related Assistance for Individuals Act of 1988. IBM-selected community service organization programs are able to purchase IBM computers at a discount for therapeutic or rehabilitative purposes.

A related IBM program is the network of IBM *Rehabilitation Training Programs*. These programs are now in over 40 locations around the country and represent a special opportunity for persons with disabilities to learn programming and other computer-operation skills.

LS&S Group, P.O. Box 673, Northbrook, IL 60065, (800) 468-4789 or (708) 498-9777. **PI, VI**

A catalog describes a range of devices for persons who are vision impaired. Products range from scanners to Braille computers, enlarging lenses to monoculars and binoculars, special lamps to mobility aids, and health aids to grippers.

Macromedia, 600 Townsend St., San Francisco, CA 94103, (800) 288-4797 or (415) 442-0200. **VI**

This company makes *MacRecorder*, a product that allows Macintosh users to record; edit; and mix voice, music, and sound effects with training materials.

Maxi Aids, 42 Executive Blvd., P.O. Box 3209, Farmingdale, NY 11735, (800) 522-6294 or (516) 752-0521 (TTY). **HI, LD, PI, SI, VI**

This company markets personal use aids for people with disabilities. Many of the devices are adaptations of equipment that nondisabled people use, such as illuminated magnifiers, talking watches and calculators, low-vision clocks and timers, kitchen tools, beeping balls, and a variety of mobility aids.

Microsystems Software, Inc., 600 Worcester Rd., Framingham, MA 01701-5342, (800) 828-2600 or (508) 626-8515 (fax). **VI, SI, PI**

This company specializes in software products appropriate for use by people with disabilities. Among its products are HandiWORD for Windows (a word-prediction program to reduce the need for keyboarding), HandiKEY for Windows (an on-screen keyboard with or without speech), HandiCHAT (a pop-up communication window involving typed and spoken text), HandiPHONE (computer-controlled, hands-free telephone operation), MAGic (an image-magnifier program), and ADAPTA-LAN (allows multiple-user access to the HandiWARE product line and MAGic software).

Nurion Industries, Station Square 3, Paoli, PA 19301, (215) 640-2345. **VI**
This firm specializes in products for blind or vision-impaired individuals. The *Laser Cane* uses light beams in forward (5 or 12 feet ahead) and height (5.5 feet above) areas to provide stimuli tactually, through the user's index finger, and auditorily, with two tones, so the user can identify the location of objects. *Polaron* is an object the user can hold or wear around the neck to detect objects in the forward and side areas. Based on an ultrasound system, it can be a primary travel aid or a secondary aid with a trained dog or cane. *Wheelchair Pathfinder* combines laser and ultrasound to help a person with vision, hearing, and mobility impairments identify objects in the forward and side areas and drop-offs of 3 inches or more.

Prentke Romich Company, 1022 Heyl Rd., Wooster, OH 44691, (800) 262-1984 or (216) 262-1984. **PI, SI**
This firm offers a wide selection of aids designed especially for persons with speech impairments and difficulty in operating switches and controls.

Raised Dot Computing, 408 South Baldwin, Madison, WI 53703, (800) 347-9594 or (608) 257-9595. **VI**
A catalog describes products such as *Hot Dots,* which converts word-processor documents to Braille output; *Flipper,* which lets you selectively access portions of the screen and have voice output; *BEX,* which translates text files to Grade I or II Braille; *MathematiX,* which allows mixed text and Nemeth-based mathematics to be presented in a form the regular teacher can use; and *pixCELLS,* which helps the user create computer graphics in brailled form. A new product, *Mega Dots,* is a Braille translation and word-processing program for the PC.

Regenesis Development Corporation, 1046 Deep Cove Rd., North Vancouver, B.C., Canada, V7G-1S3, (604) 929-6663. **PI**
The *Robot Manipulator* is a table-top device that is voice controlled through a computer to direct a robot arm in grasping and manipulating objects. It is fully programmable in up-down, left-right, and rotate movements.

Tech Aid, Inc., Ste. 198, 5464 N. Port Washington Rd., Milwaukee, WI 53217, (800) 451-2773. **PI**
Tech Aid is the supplier for the *Lightwriter* family of devices, communication aids which vary in input from full keyboard to foot, puff (breath), and eye-blink control of text materials. The machines have attachments for external printers and telephone use.

Telesensory Systems Inc., 455 N. Bernardo Ave., Mountain View, CA 94043-5274, (800) 227-8418. **VI**   Among the devices available are the following:

Image enlargers: *Vantage* and *Voyager* magnify text or shallow objects 45 to 60 times the original size; include black and white images, brightness and contrast control, and image reversal; *Chroma* magnifies up to 60 times, in color.

Computer display enlargers: *Vista* is a mouse-controlled computer screen-enlarging system; *Lynx* links image magnifiers and computers; *DP-11 Plus* magnifies text for IBM computers.

Speech systems: *Vert Plus* and *Personal Vert* convert text in computer software to speech, accessed by earphone, at rates of 300 to 600 WPM.

Optical scanning system: *OsCar* scans documents and translates text to Grade 2 Braille and places it in a file for word processing (e.g., Word Perfect).

Optacon reading device: *Optacon II* is a portable text-to-tactile image device for reading ink print directly. With attachments, it can access computer text via *Optacon PC* for the IBM and *inTouch* (from Berkeley Systems) for the Macintosh.

Braille computer systems: *Navigator,* when coupled with a computer, provides a user with a Braille equivalent of what is on the computer screen; *BrailleMate,* a pocket computer, accepts Braille input (such as note taking in class) and automatically converts it to standard text for speech and print output; *VersaPoint* provides Braille embossing from computer output, creates tactile graphics generator with *VersaPoint Graphics* or *Illustrations* (by Lorin Software), and prints arithmetic with *Braille Arithmetic; MPrint,* an attachment for a standard Braille printer, can produce inkprint output with external printer.

Telecommunications device: *TeleBraille II* allows face-to-face communication with a deaf-blind person or telephone communication (Braille input, standard text output) with another person. It is used with a TDD (Telecommunications Device for the Deaf) device available from other manufacturers.

Toys for Special Children, 385 Warburton Ave., Hastings-on-Hudson, NY 10706, (800) 832-8697. **HI, LD, PI, VI**

A catalog describes products for very young children, including a wide range of specially created switches and mounting hardware for adapted toys.

Vysion, Inc., 30777 Schoolcraft, Livonia, MI 48150-2010, (313) 522-3300. **LD, SI, VI**

Personal Speech System is a self-contained text-to-speech synthesizer that operates as a peripheral to a variety of computers. In addition to generating speech, the Personal Speech System provides comprehensive sound effects and music at moderate cost. An additional product, Type-'N-Talk, allows easy programming of speech into a computer, with an unlimited vocabulary.

Xerox Imaging Systems, 9 Centennial Dr., Peabody, MA 01960, (800) 248-6550 or (508) 977-2000. **LD, VI**

This is technology (developed by Kurzweil Personal Reader) for reading machines that convert text to speech or to a screen display. The company also offers *Book Wise* for dyslexic students.

ZYGO Industries, Inc., P.O. Box 1008, Portland, OR 97207-1008, (800) 234-6006 or (503) 684-6006. **PI, SI, VI**

ZYGO markets a variety of communication and technical aids. Included in their catalog are the *Macaw II,* a digital recording unit for speech output, the *QED Scribe Communication Aid,* the PACA *Portable Anticipatory Communication Aid* (software only), and the *Tetra Scan II* auxiliary keyboard for computers (to be used by quadriplegic persons who cannot manage a standard keyboard).

DISABILITY-ORIENTED ORGANIZATIONS

Alexander Graham Bell Association for the Deaf, 3417 Volta Place, NW, Washington, DC 20007, (202) 337-5220 (Voice and TDD).

This association is a leader in promoting the auditory/oral (speechreading) methods of communication between deaf and hearing individuals. It handles over 20,000 inquiries a year, dealing with such topics as teacher training, oral interpreting services, lip-reading courses, television captioning, and signaling devices. The Oral Hearing Impaired Section (OHIS) is a network of hearing-impaired adults who choose to communicate through spoken language and speechreading and serve as models for children.

American Association for the Advancement of Science, 1333 H Street, NW, Washington, DC 20005, (202) 326-6630 (voice or TDD). Contact: Director, Project on Science, Technology, and Disability.

Through its Project on Science, Technology, and Disability, AAAS provides an information center to link people with disabilities and their families, with scientists and engineers with disabilities who are willing to share their coping strategies in education and career advancement. Their project Science-by-Mail links students and scientists as "pen pals."

AAAS has published a series of four booklets called *Barrier-Free in Brief.* One addresses accommodating needs in the classroom (*Laboratories and Classrooms in Science and Engineering*); another addresses settings in out-of-school programs (*Access to Science Literacy*); a third (*Access in Word and Deed*) provides a list of eighty consultants who have agreed to serve as informal consultants on questions about accommodations and assistive technology; the fourth deals with *Workshops and Conferences for Scientists and Engineers.*

Booklets entitled *Find Your Future* and *You're in Charge* can be obtained from AAAS and are for secondary students with potential in science. *Find Your Future* will help such students decide if a science career is appropriate for them; *You're in Charge* will help them prepare for post-secondary entry.

American Foundation for the Blind, 15 West 16th St., New York, NY 10011, (800) 232-5463 or (212) 620-2080.

This organization operates a technology center, provides research on new devices, and provides an information system and catalog concerning adaptive devices for the blind and vision impaired. The catalog includes an audible carpenter's level, a liquid level detector, talking calculators, TeleBraille devices, tactile maps, tactile measures, rules, micrometers, talking scale, audible battery testers.

American Paralysis Association, c/o Montebello Rehabilitation Hospital, 2201 Argonne Dr., Baltimore, MD 21218, (800) 526-3456.

This organization operates the Spinal Cord Injury Hotline information and referral service.

American Speech, Language and Hearing Association, 10801 Rockville Pike, Rockville, MD 20852, (800) 638-8255 or (301) 897-5700.

This association provides an information center for concerns in the area of speech pathology and audiology.

Association on Higher Education and Disability, P.O. Box 21192, Columbus, OH 43221, (614) 488-4972 (voice and TDD).

AHEAD provides information about disability support service offices for students in over 600 higher education institutions and answers questions about legal rights and testing accommodations for students with disabilities.

Association for Retarded Citizens, 500 East Border St., Ste. 300, Arlington, TX 76010, (817) 261-6003.

ARC maintains a large information system about the needs and services available for people with severe disabilities, including persons who are retarded.

Cerebral Palsy Research Foundation, 2021 North Old Manor, P.O. Box 8217, Wichita, KS 67208, (316) 688-1888.

This foundation conducts research on new assistive devices and maintains information about a wide variety of technology.

Council for Exceptional Children, 1920 Association Drive, Reston, VA 22091-1589, (703) 620-3660 (voice and TDD).

CEC is an organization for educators of persons with disabilities, including support personnel, researchers, and families. It provides information services and operates the ERIC Center containing numerous documents related to disability.

Epilepsy Foundation of America, 4351 Garden City Dr., Landover, MD 20785, (301) 459-3700.

The foundation provides information about epilepsy, the needs of individuals who have it, and the services available for them.

Gallaudet University, 800 Florida Ave., NE, Washington, DC 20002, (202) 651-5257 or (202) 672-6720 (voice and TDD).

Gallaudet operates a National Information Center on Deafness and also offers a Technology Assessment Program, which develops and evaluates technology for persons who are speech impaired and hearing impaired or deaf.

HEATH Resource Center (Higher Education and Adult Training for People with Disabilities), Ste. 800, One Dupont Circle, Washington, DC 20036, (800) 544-3284 (voice and TDD) or (202) 939-9320 (voice and TDD).

HEATH maintains lists of colleges and universities providing access and services for students with disabilities.

Job Accommodation Network, West Virginia University, 809 Allen Hall, Morgantown, WV 26506, (800) 526-7234 or (304) 293-7186.

JAN provides information about specific problems in making a laboratory accessible for persons with different disabilities. (Be prepared to discuss the specific circumstances of the case.)

Learning How, P.O. Box 35481, Charlotte, NC 28235, (704) 376-4735.

This group has chapters around the U.S. that emphasize mentor programs. Role models from the community are provided for persons with disabilities.

National Association for the Visually Handicapped, 22 West 21st St., New York, NY 10010, (212) 889-3141.

This association provides publications and products designed for use with persons who have vision impairments.

National Braille Association, Inc., 1290 University Ave., Rochester, NY 14607, (716) 473-0900.

As a service to vision-impaired readers, this association maintains a depository for hand-transcribed Braille masters, and Braille copies of items in the collection are sold below cost. The approximately 1,800 titles are listed in four catalogs: college and graduate level textbooks, music, general interest, and technical tables bank. The latter includes science and mathematics tables found in high school and college textbooks.

National Captioning Institute, 5203 Leesburg Pike, Falls Church, VA 22041, (703) 998-2400 (voice or TDD).

NCI is a nonprofit organization and a leader in bringing closed-captioning to television. In addition to their role as a captioning service, NCI manufactures and distributes the *TeleCaption Decoder.* They have also developed *Audio-Link,* a wireless audio system for hard-of-hearing people to better enjoy public events.

National Center for Learning Disabilities, 99 Park Ave., New York, NY 10016, (212) 687-7211 or (703) 451-2078.

This center offers information and referral services for persons with learning disabilities.

National Easter Seal Society, 70 East Lake St., Chicago, IL 60601, (312) 726-6200 (voice) or (312) 726-4258 (TDD).

This organization is a national network of affiliated service agencies throughout the U.S. They share one purpose: to help people with disabilities achieve maximum independence. Easter Seals provides rehabilitation services, technological assistance, prevention services, advocacy, and public education programs.

National Federation of the Blind, 1800 Johnson St., Baltimore, MD 21230, (301) 659-9314.

This federation provides information about adaptive technology for the blind.

National Registry of Interpreters for the Deaf, 8719 Colesville Rd., Ste. 310, Silver Spring, MD 20910, (301) 608-0050 (voice and TDD).

This registry provides information about local interpreter services.

National Rehabilitation Association, 633 South Washington St., Alexandria, VA 22302, (703) 836-0850 (voice) or (703) 836-0852 (TDD).

The association provides information about rehabilitation services around the country.

National Rehabilitation Hospital, 102 Irving St., Washington, DC 20010, (202) 877-1932.

Research on the effectiveness of assistive and adaptive technology is conducted in the Rehabilitation Engineering Program at the hospital.

National Rehabilitation Information Center and ABLEDATA, 8455 Colesville Rd., Silver Spring, MD 20910, (800) 346-2742 (voice or TDD) or (301) 588-9284 (voice or TDD).

The center operates a database on rehabilitation information as well as ABLEDATA and will conduct literature searches on specific topics.

National Technical Institute for the Deaf, P. O. Box 9887, Rochester, NY 14623-0887, (716) 475-6400 (voice) or (716) 475-2181 (TDD).

NTID serves about 1,100 persons with disabilities from all around the U.S. and is affiliated with the Rochester Institute of Technology, which serves approximately 12,000 nondisabled students. NTID prepares persons with speech and hearing disabilities for employment (96 percent placement rate). This success is due in part to their varied career emphases, their well-developed cooperative education program with employers, and their National Center for Employment of the Deaf, which conducts seminars and training sessions for employers of the deaf.

Paralyzed Veterans of America, 801 18th St. NW, Washington, DC 20006, (202) 872-1300.

This organization publishes a journal and other information pertaining to rehabilitation devices for use by people with disabilities. Their information and services are not restricted to veterans or a particular age group.

Recording for the Blind, 20 Roszel Rd., Princeton, NJ 08540, (800) 221-4792 or (609) 452-0606.

This organization has chapters throughout the country where 4,800 people voluntarily prepare textbooks, fiction, and other educational material on recorded tapes. Although the organization's name suggests a service for the blind, more than half of the 27,000 users of the service are learning disabled.

Self-Help for Hard-of-Hearing People, Inc., 7800 Wisconsin Ave., Bethesda, MD 20814, (301) 657-2248 (voice) or (301) 657-2249 (TDD).

SHHH consists of 25,000 volunteers and has local chapters in every state. Services include a bimonthly magazine called the *SHHH Journal,* referral and advisory services, an information and resource center, special publications and programs, national conventions, and local group meetings for mutual support. Small annual membership fee.

Sensory Access Foundation, 399 Sherman Ave., Palo Alto, CA 94304, (415) 329-0430.

The foundation provides information and assistance in acquiring adaptive technology for persons with sensory impairments.

Stanford Children's Hospital Rehabilitation Engineering Center, 520 Sand Hill Rd., Palo Alto, CA 94304, (415) 324-9991.

This center provides information and assistance in selecting and constructing adaptive equipment for children with disabilities, particularly communication aids and prosthetics for children with severe physical impairments.

Technical Assistance Resource Center, 1101 Connecticut Ave., NW, Washington, DC 20036, (202) 857-1140 (voice and TDD).

This center maintains a referral network related to the legislation known as the Technology-Related Assistance for Disabilities Act of 1988. It has information on various technologies, such as page turners, wheelchairs, and computers.

Telecommunications for the Deaf, 8719 Colesville Rd., Ste. 300, Silver Spring, MD 20910, (301) 589-3786 (voice) or 589-3006 (TDD).

This organization promotes the use of TDDs and other adaptive aids for the hearing impaired, including text telephones and telecaptioning. It produces the *International Directory of TDD Users,* a telephone directory of TDD numbers and companies providing services for the deaf.

The Association for Persons with Severe Handicaps (TASH), 11201 Greenwood Ave. North, Seattle, WA 98133, (206) 361-8870.

This association provides information about assistance to persons with severe disabilities.

Trace Research and Development Center, Waisman Center, University of Wisconsin-Madison, 1500 Highland Ave., Madison, WI 53705, (608) 262-6966 (voice) or (608) 263-5408 (TDD).

Trace operates the Rehabilitation Engineering Center on Access to Computers and Electronic Equipment and is one of the leading organizations for the development, tailoring, and evaluation of computers and adaptive devices for people with disabilities. Trace has a program focused on high-technology communication aids enabling nonspeaking and physically impaired persons to converse and write and a program on control mechanisms, including home environment controls. Trace has developed *HyperABLEDATA,* a microcomputer version of the on-line database of over 17,000 products maintained by the

Newington Children's Hospital, Newington, CT. *HyperABLEDATA* is available on CD-ROM for use with Macintosh and, soon, IBM computers.

United Cerebral Palsy Foundation, 1522 K St., NW, Washington, DC 20005, (202) 842-1266.

The foundation conducts research projects, provides information on technology, and promotes the interests of persons with cerebral palsy.

SCIENCE AND TECHNOLOGY RESOURCE ORGANIZATIONS

Association of Science-Technology Centers, 1413 K St., NW, Washington, DC 20005-3405, (202) 371-1171.

This nonprofit organization of museums and affiliated institutions is dedicated to increasing public understanding of science and technology. The association organizes and circulates hands-on exhibitions to museums throughout North America. Literature from the Association that identifies and locates more than 160 science centers throughout the U.S. is available.

Center for Multisensory Learning, Lawrence Hall of Science, University of California, Berkeley, CA 94720, (510) 642-8941.

The Lawrence Hall of Science is deeply involved in training public school teachers in the use of hands-on science techniques in regular classrooms. They have developed curriculum materials with special emphasis on making them appropriate for students with disabilities:

Science Activities for the Visually Impaired/Science Enrichment for Learners with Physical Handicaps (SAVI/SELPH). Modules were developed for hands-on use by students with disabilities in grades 4–7. Multisensory investigations are presented in life, earth, and physical science in nine areas: communication, environments, environmental energy, kitchen interactions, magnetism and electricity, measurement, mixtures and solutions, scientific reasoning, and structures of life.

Full Option Science System (FOSS). These curriculum materials are an extension of the SAVI/SELPH materials. While they were created for regular students, mention is made within the modules about adaptations that can be made for different disabilities. Topics are drawn from the physical, life, and earth sciences, and scientific reasoning and technology and are correlated with leading texts. Two modules are at the pre-K and K level, six for grades 1 and 2, eight for grades 3 and 4, and ten for grades 5 and 6.

Modern Talking Picture Service, 5000 Park St. North, St. Petersburg, FL 33709, (800) 237-6213 (voice and TDD).

This service rents a number of captioned films.

NADA Scientific Ltd., P.O. Box 1336, Champlain, NY 12919-1336, (800) 233-5381.

NADA is one of a number of firms supplying school-level science equipment for demonstration and student use and is the authorized U.S. distributor for Nakamura Scientific Co., Ltd. of Tokyo. Their catalog organizes the equipment around science content (microscopes, electricity, electrostatics, magnetism, dynamics, waves, light, heat, sound, liquid and pressure, chemistry, earth science, and measurement). A number of the devices are appropriate for, or can be adapted for, use by disabled persons.

National Geographic Society, Washington, D.C. 20077-9966, (800) 368-2728.
The Society operates the National Geographic *Kids Network* (developed by TERC), which allows students in different schools and states to hook up through telecommunications and share their hands-on experiences in science. Data developed in one location can be transmitted and used with low-cost computer equipment that students in grades 4–6 can operate. Courses available through the network cover science, computer science, statistics, math, geography, language arts, and social studies. The society also makes available other computer software (separate from the network) pertaining to science topics in the elementary and middle schools, such as weather, taught with real data.

National Science Teachers Association, 1742 Connecticut Ave., NW, Washington, DC 20009-1171, (202) 328-5800.
With over 50,000 members, NSTA is the leading professional organization for science education. Its publications include *Science and Children, Science Scope, The Science Teacher,* and the *Journal of Science Teaching,* and documents on particular issues and topics. Conventions are held annually at regional and national levels with extensive exhibits and vendor displays, as well as instructional programs, many of which involve practical how-to demonstrations.

Optical Data Corporation, 30 Technology Dr., Warren, NJ 07059, (908) 668-0022.
Windows on Science is a videodisc-based primary science curriculum containing still images, movie clips, and two audio tracks with separate English and Spanish narrations. The program contains material suitable for life, earth, and physical science study in grades 1–3. Also available is *The Living Textbook.*

Technical Education Resource Center (TERC), 2067 Massachusetts Ave., Cambridge, MA 02140, (617) 547-0430.
TERC is a research and development firm in the area of science and mathematics education. They enjoy a reputation as pioneers in innovative, technology-oriented, hands-on science curriculum development. Their recent development work includes a) an emphasis on global issues (such as acid rain), b) activities that engage students interactively through microcomputer networks such as the National Geographic *Kids Network,* c) a videocassette on the TERC Star Schools curriculum, and d) a series of experiments in different science disciplines. TERC publishes a semi-annual newsletter *Hands On!* and has written *Beyond Drill and Practice: Expanding the Computer Mainstream,* available from the Council for Exceptional Children, 1920 Association Dr., Reston, VA 22091; call (703) 620-3660.

Videodiscovery Inc., 1700 Westlake Ave. North #600, Seattle, WA 98109, (800) 548-3472 or (206) 285-5400.
Videodiscovery is one of several firms engaged in supplying new videodisc technology for school use. These videodiscs are rich sources of visual material about nature that cannot be readily accessed locally or reproduced as comprehensively in printed formats. Videodiscovery also supplies Hypercard stacks for use with Macintosh computers and *Vidkit* authoring software that allows schools to tailor science instruction from extensive graphic materials. In connection with the biological science disc, for example, stacks are available dealing with life cycles and cell biology. Discs are available in chemistry, physics, evolution, life cycles, earth science, and additionally, *Atoms-to-Anatomy* and *Science Discovery: Science Sleuths.*

Guidelines for an Inservice Workshop

Given here are guidelines for presenting an inservice workshop on how to include students with disabilities in science classes.

1. The purposes of the workshop should be made known to all. Several purposes seem apparent; you may choose to add more.

   • The workshop is an opportunity for teachers to develop their skills in an area (including K–12 students with various disabilities) and instruction in a subject matter (science).

   • The workshop is an opportunity for school staff to share expertise and ideas of common concern, whether they are regular classroom teachers, special education teachers, technical specialists, or administrators.

   • The workshop can provide everyone with a sense of working together with the same goal in mind—quality education for underserved students with disabilities who have much to contribute in society and in the school.

   • Individual students with disabilities will benefit from the planning and instruction that will result from the workshop.

2. The following steps have been used to conduct successful inservice workshops across the country.

   a. Identify a date for the workshop and publicize it reasonably far in advance. Put one person in charge and reserve places in the order of response until the seating capacity of the meeting room is reached.

   b. Arrange for the formation of teams of teachers that represent the different grade levels and schools in the district. The focus should be on science teachers who teach regular classes. A 6:1 ratio of regular teachers to special education teachers attending the workshop can be effective.

   c. Provide an incentive to attend. Past workshops have chosen a number of different incentives, including a free copy of *Science Success* (purchased by the district) for future reference use, a small stipend, release time, free refreshments or a free meal, and continuing education credits.

   d. Have a sufficient number of copies of *Science Success* available so that each teacher has one to use. Any additional materials to be shared should be duplicated in advance so they can be passed out at the beginning of the workshop.

e. Allow about three hours for the workshop. Take a break after an hour and a half or so.

f. Share the presentation between two leaders, each of whom should be thoroughly familiar with the contents of *Science Success*. At least one leader should have experience in teaching students with disabilities in a science class. Enthusiasm and sincerity are a must.

g. Provide for some type of follow-up evaluation to determine whether the teachers did, in fact, implement the things they learned in the workshop.

3. Actively involve teachers in the conduct of the workshop. The procedures suggested below have been well received by teachers.

a. Welcome everyone. In a few opening remarks, explain the purpose for the workshop and express appreciation for those who chose to attend. Make the point that students with disabilities are, first and foremost, students with much in common with their peers, but with some unique needs. Give an example or two from experience to set the group at ease. Explain that those in attendance are leaders who value these students and are looking for ways to improve their own teaching. Allow 5 minutes.

b. Ask each of the teachers to identify themselves and their schools, and to tell in a sentence what they hope to learn in the workshop and the extent of their prior experience with students with disabilities. Depending on the size of the group, allow 5 to 10 minutes.

c. Break the group into teams of four to six persons, with at least one team assigned to each of the six disabilities (vision impairment, hearing impairment, physical impairment, speech impairment, emotional disorder, learning disability). Allow 10 to 15 minutes, and conduct the role-playing exercise in the Overview section of *Science Success*.

Allow about 3 minutes per group for an oral summary to the main group. Have a team representative report what disability (or disabilities) the team considered and what they noted about (a) difficulties encountered in the morning routine and (b) difficulties encountered in the classroom situation. Ask each teacher who knows of a student with a disability to write down a brief description of the disability and what, in their view, is particularly challenging about the case from an instructional point of view. Allow 5 minutes, and collect the descriptions.

d. In a leader-conducted session, review the table of contents, then divide the room into four sections. Ask the first section to examine Chapter 1 of *Science Success* with the goal of identifying key ideas they think should be pointed out to the other two groups. At the same time, have the second section look at Chapter 2, the third section look at Chapter 3, and the fourth section look at Chapter 5 (skip Chapter 4 for this exercise). Allow 10 minutes for the examination review process.

Use an overhead transparency or chalkboard to summarize the key ideas that are voluntarily reported from members of section 1, followed by section 2, section 3, and section 4. Acknowledge and reaffirm the important points presented. Allow 5 minutes for each group report.

Elapsed time: 80 minutes.

e. Take a 10-minute break. During the break, the workshop leader(s) should look through the case descriptions turned in by the teachers, select two or three that present different kinds of challenges to the teacher, and set them aside for use in the second session.

f. In a leader-conducted session lasting 10 to 15 minutes, go over the introduction to Chapter 4 so everyone understands that it can be accessed like a reference section. Tell them you will be giving them an assignment that calls for using this chapter. Be sure that everyone understands the symbols (see page 38), the structure of the chapter, and how to use the chart on page 38 effectively. Make clear that the chapter contains the core of the alternative strategies recommended for each disability. If the leader senses that there are still questions about how to use the chapter, walk the group through the first topic.

Have the original small groups reconvene; each group should be focused on one disability. Have them read the case example that applies to that disability (Appendix C). Have them elect a reporter (or two) for the group and give that individual a clear overhead transparency and a pen, with instructions to prepare a list of strategies suitable for the case that will be shared with the whole group. Allow 20 minutes for the teams to come up with a comprehensive plan for serving that student. (While discussion proceeds, monitor the group reporters to be sure strategies are being recorded on the transparency.)

When the discussion time is up, have each group reporter present the findings to the group. Stimulate questions, comments, and additional ideas from the group. Allow 5 minutes per team for the presentation and discussion.

g. Using the two or three descriptions of real cases selected previously, conduct a general session in which each case is read aloud and the teachers volunteer ideas for how they would approach it. The idea is that when teachers came to the workshop they may not have had a clear plan for working with students with disabilities; after studying *Science Success* they are now being asked to serve as a mutual-support resource team. Teachers' ideas should generally reflect what they have read in *Science Success* or learned in the discussions in the workshop. Allow 15 to 20 minutes for the overall discussion.

h. Close the second session. Remind teachers that the student with a disability also has *abilities* and can do many things in the science class if given the appropriate support and encouragement. Each individual or team returning to its school has a dual mission to fulfill: (a) sharing their workshop findings and *Science Success* with other teachers and (b) improving educational opportunities for students with disabilities in their science classes (and other classes) in the school.

If an evaluation follow-up is to be done after the workshop, describe the procedure. Thank everyone for their participation. Allow 5 minutes.

Elapsed time, including the break and session two: 165 minutes (overall, 3 hours).

Guidelines for a Two-Session Preservice Class

1. The purpose of the class should be made known to all. Several purposes seem apparent; you may choose to add more.

   - The class is designed to prepare students for successfully working as teachers with students with disabilities.

   - The class builds a sense of awareness and understanding of disability-related issues and, through the use of *Science Success*, informs them about strategies for effective teaching.

2. The following steps have been used to conduct successful classes in science education at multiple colleges and universities:

   a. In advance of the date of the first class on the topic, give the students a homework assignment. The assignment is to carry out the role-playing activity described in the Overview section of *Science Success* and to prepare and orally report on the problem situations identified. (Presumably, *Science Success* has been named as a supplementary text for the course and most students will have obtained it at the bookstore.)

   b. In rotation across the disabilities, have selected students report on what they identified as problems. Allow 10 minutes. Summarize and distinguish the various kinds of barriers encountered: attitudinal, environmental, and personal.

   c. Break the class into four groups. Ask the first group to examine Chapter 1 with the goal of identifying key ideas they think should be pointed out to the other groups. At the same time, have the second group look at Chapter 2, the third look at Chapter 3, and the fourth look at Chapter 5. Allow 10 minutes for the examination review process. Allow 10 more minutes for summary presentations by leaders chosen by each group.

   d. In an instructor-led session lasting 10 to 15 minutes, go over the introduction to Chapter 4 so everyone understands that it can be accessed like a reference section. Tell the class that you will be giving them an assignment that calls for using this chapter. Be sure that everyone understands the symbols (see page 38), the structure of the chapter, and how to use the chart on page 38 effectively. Make clear that this chapter contains the core of the alternative strategies recommended for each disability. If the leader senses that there is any question remaining about how to use the chapter, walk the group through the first topic.

e. Taking about 5 minutes, give the homework assignment. Randomly assign the six case vignettes in Appendix C to the class. Have them use *Science Success* as the basis for preparing an intervention plan for the student with a disability. The intervention plan is to contain

- A description of strategies appropriate for the student in (a) teacher-directed instruction, (b) student-centered instruction, and (c) instruction involving partners and team cooperation

- A description of the resources (see Chapter 5) that might be helpful for the student and supportive of the instructional plan

- A description of what other school-level resources (people and technology) they might draw upon and how they would use them

Elapsed time: 50 minutes.

f. In the second session, organize the class into six groups, by disability. Allow 20 minutes for them to interact and to share their various instructional intervention plans. Have them choose a group leader, arrive at a group consensus about the proposed strategies, and list them on an overhead transparency or two. Have the group leaders present them to the class. Collect the homework.

g. Acknowledge the thoughtfulness and thoroughness of the group plans (applause is often appropriate). Add examples from the instructor's experience to enrich the discussion or to further make the argument for important omissions that might have been noted. For example, did the group indicate in its plan that the student with a disability would be involved in planning solutions to barriers? Allow 5 to 10 minutes.

h. Allow 20 minutes for a lecture/discussion on the following points.

- *Providing access and instruction that accommodates students with disabilities is both socially responsible and mandated under the law.* Review the key points from current legislation, including the Americans with Disabilities Act (ADA), currently applicable Federal law governing the Department of Education and the provision of special education services at the local level, the Rehabilitation Act as amended, and any relevant state legislation. A handout covering these points is suggested.

- *Affirmative action is everyone's job.* If a school district tries to evade its responsibilities by not supplying appropriate support to the student (e.g., no interpreter is provided when needed or no needed adaptive equipment is purchased), it is a joint responsibility of the instructional staff, the family, and the student to stand up for the student's rights under the law.

i. Arrange a visit to the class by a successful adult with a disability, preferably in a professional or technical field associated with science. For the balance of the time, approximately 20 minutes, and following an appropriate introduction, hold a question-and-answer session between the guest and the class.

j. Close with statements of optimism about the teacher-trainees' future interactions with students with disabilities in their classes.

Elapsed time: 50 minutes (overall time 100 minutes).

Case Vignettes

PHYSICALLY IMPAIRED

José is 10 years old and in fourth grade. When he was three, he was in an automobile accident. As a result his legs became paralyzed and his arm movements somewhat restricted. He uses a motorized wheelchair. According to his school records, he is shy by nature and reluctant to volunteer during class activities. José will enter your class from an out-of-state public school in November, when your class will be doing a unit on weather and seasonal changes.

1. What steps can you take to ensure José gets off to a good start in your class and participates fully?

2. What sort of help is José most likely to need from support staff or other students?

3. If you were planning an assignment for the class to bring in examples of how living organisms prepare for seasonal changes (e.g., objects, leaves, pictures), how might you modify this assignment for José?

4. Students in the class will be obtaining materials and specimens for seasonal changes in living things. What strategy might you use to enable José to participate in this activity?

5. This science unit will include computer work. José has used a computer in his other school. What might you need to check to be sure José can participate?

VISION IMPAIRED

Janet is a 16-year-old high school sophomore. Legally blind since birth, she can only see in a narrow field and nonglare lighting is critical. She reads Braille and uses a cane. She is especially interested in music. Her parents are very supportive. Janet is transferring from a neighboring public school. Her school records indicate that she is on the quiet side and often keeps to herself, though she responds well on a one-to-one basis. When she joins your class, you will be doing a unit on the structure and function of living cells.

1. What steps can you take to ensure Janet gets off to a good start in your class and participates fully?

2. What sort of help is Janet most likely to need from support staff or other students?

3. If you were planning to show slides of cell structures during a lecture in the classroom, how might you modify the lesson to make it more meaningful for Janet?

4. What arrangements might you make so Janet can find and use equipment in the science center?

5. You are organizing a cooperative-learning activity related to cell reproduction. What issues do you need to consider in assigning Janet to a group and giving her a task to complete?

HEARING IMPAIRED

Kim is an 11-year-old fourth grader with impaired hearing. He wears aids in both ears and lip-reads fairly well. He has been attending a community school for the deaf, but his parents want him to attend public school. He knows how to sign but his peers at the new school don't sign. Teachers at the community school believe that he can make the transition. Kim is not sure: he is scared about attending a new school and interacting with hearing students. When he arrives in your class, you will be doing a unit on sources and composition of light.

1. What steps can you take to ensure Kim gets off to a good start in your class and participates fully?

2. What sort of help is Kim most likely to need from support staff or other students?

3. In the past you have used oral presentations (for information sharing combined with oral student reports) to introduce and summarize topics. How might you use other modalities so Kim can be sure to get the essential points made in these presentations?

4. In the science center, Kim may not be able to see you well enough to lip-read and may get lost in the middle of some instructions you are giving. What system might you devise so that he can communicate his needs for repetition or clarification?

5. You are planning to have students work in groups to develop models that demonstrate the effects of lunar and solar eclipses. What special steps might you need to take to ensure that Kim can participate fully in his group?

LEARNING DISABILITY

Michael, a 13-year-old sixth grader, has a diagnosed reading disability. He is energetic, prone to burst out with comments during class, and enjoys being the class clown. He demands lots of individual attention, avoids anything involving reading whenever he can, and fails to complete homework. However, he is good with mechanical things and can draw well. He is transferring into your class because of staff reductions. You are getting ready to do a unit on sound.

1. What steps can you take to ensure Michael gets off to a good start in your class and participates fully?

2. Is there something you can do to interest him in science and head off his tendency to avoid homework?

3. You intend to give the class a fairly complex assignment during this unit. What steps might you take to be sure Michael pays attention until he understands what is to be done?

4. One activity in this unit will take place in the science center and another in the library. You are concerned that Michael may get off the track during these activities, paying more attention to the work of others than to his own work. What strategies might keep him on track?

5. Students in your class are to work in groups of three to create a homemade musical instrument and then show it to the class. You know from Michael's previous teacher that students resist being teamed up with him because he tends not to do his share of the work. What might you do to help his group work productively?

SPEECH IMPAIRED

Kristin, an 8-year-old third grader, has a cleft palate. She is scheduled to be in your class when school resumes in the fall. She has been receiving speech instruction from district staff, but her speech is still very difficult to understand. According to school records, she gets frustrated when she cannot communicate and is prone to emotional outbursts to vent her frustration. She has few friends. Other students do not choose to work or play with her, though she continues to seek their companionship, and they view her as an outsider. Kristin often sits by herself and draws pictures, and they reveal talent. You will begin the class with a unit on earth science and a focus on rocks.

1. What steps can you take to ensure Kristin gets off to a good start in your class and participates fully?

2. For one of the activities in this unit, each student is to bring a small rock to class, show it to the class, and describe it in terms of the characteristics learned in the unit. What support from others might Kristin use to carry out this assignment?

3. You know that Kristin tries to avoid talking whenever she can, and you want to be sure she communicates her needs to you. What strategies might you use to ensure that she lets you know when she doesn't understand or needs help?

4. Kristin has been playing with the computer at the speech therapist's office. How might you allow her to harness this emerging interest in your science class? Are there other technologies she might benefit from using?

5. The final project in your unit is a team activity, with oral reports from each team member. How will you decide which team Kristin should be on and what role she should take on that team?

EMOTIONAL DISORDER

Liza is a 12-year-old fifth grader who has attended a number of schools. The various records indicate that she has difficulty focusing on class activities and has poor study habits. Her reading test scores indicate high academic potential, but her performance is lower than average. She may have been subjected to abuse as a young child. Her mother has given her up, and Liza has lived in several foster homes. Peers are very important to her, but rather than work at being friendly she is unsociable and tends to hit and shove when provoked. She likes small animals. She is being placed in a new home in your community and will enter your class just as you begin a unit on electricity and magnetism.

1. What steps can you take to ensure Liza gets off to a good start in your class and participates fully? Can you think of a way to combine her interest in small animals with a unit on magnetism?

2. A parent has offered to come into your class each day; it happens to be during the time that you teach science. How might you use this volunteer to help Liza?

3. About a month after Liza enters the class, a field trip to a science museum is scheduled. In addition to making the usual arrangements for a field trip, what preparations might you make to ensure that this is a good experience for Liza?

4. Another activity will require your students follow fairly complex directions to construct and test an electric circuit. What strategies might you use to be sure that Liza (and other students) understand the directions you will give for this activity?

5. Liza may not want to call attention to herself by asking for help. How will you know when she needs help? How can you help her communicate her needs?